Ergebnisse der Mathematik und ihrer Grenzgebiete 14

A Series of Modern Surveys in Mathematics

H. S. M. Coxeter W. O. J. Moser

Generators and Relations for Discrete Groups

Fourth Edition

With 54 Figures

Springer-Verlag
Berlin Heidelberg New York 1980

H. S. M. Coxeter

Department of Mathematics
University of Toronto
Toronto M5S 1A1, Ontario
Canada

W. O. J. Moser

Department of Mathematics
McGill University
Burnside Hall
805 Sherbrooke St.
Montreal H3A 2K6, Quebec
Canada

AMS Subject Classifications (1970): 05 B 45, 05 C 10, 05 C 20, 05 C 25, 55 A 05, 50 C 15, 20 E 40, 20 F 05, 20 F 10, 20 F 15, 20 H 05, 20 H 15, 20 H 20, 20 K 05, 20 G 40

ISBN 3-540-09212-9 Springer-Verlag Berlin Heidelberg New York
ISBN 0-387-09212-9 Springer-Verlag New York Heidelberg Berlin

ISBN 3-540-05837-0 3. Auflage Springer-Verlag Berlin Heidelberg New York
ISBN 0-387-05837-0 3rd edition Springer-Verlag New York Heidelberg Berlin

Library of Congress Cataloging in Publication Data. Coxeter, Harold Scott Macdonald, 1907—. Generators and relations for discrete groups. (Ergebnisse der Mathematik und ihrer Grenzgebiete; Bd. 14). Bibliography: p. Includes index. 1. Discrete groups. 2. Groups, Theory of—Generators. 3. Groups, Theory of—Relations. I. Moser, W.O.J., joint author. II. Title. III. Series. QA171.C7. 1979. 512′.2. 79-23198

Printing and binding: Brühlsche Universitätsdruckerei, Gießen
2141/3140-543210

Preface to the First Edition

When we began to consider the scope of this book, we envisaged a catalogue supplying at least one abstract definition for any finitely-generated group that the reader might propose. But we soon realized that more or less arbitrary restrictions are necessary, because interesting groups are so numerous. For permutation groups of degree 8 or less (i.e., subgroups of $\mathfrak{S}_8$), the reader cannot do better than consult the tables of JOSEPHINE BURNS (1915), while keeping an eye open for misprints. Our own tables (on pages 134—142) deal with groups of low order, finite and infinite groups of congruent transformations, symmetric and alternating groups, linear fractional groups, and groups generated by reflections in real Euclidean space of any number of dimensions.

The best substitute for a more extensive catalogue is the description (in Chapter 2) of a method whereby the reader can easily work out his own abstract definition for almost any given finite group. This method is sufficiently mechanical for the use of an electronic computer.

There is also a topological method (Chapter 3), suitable not only for groups of low order but also for some infinite groups. This involves choosing a set of generators, constructing a certain graph (the Cayley diagram or DEHN*sche Gruppenbild*), and embedding the graph into a surface. Cases in which the surface is a sphere or a plane are described in Chapter 4, where we obtain algebraically, and verify topologically, an abstract definition for each of the 17 space groups of two-dimensional crystallography.

In Chapter 5, the fundamental groups of multiply-connected surfaces are exhibited as symmetry groups in the hyperbolic plane, the generators being translations or glide-reflections according as the surface is orientable or non-orientable.

The next two chapters deal with special groups that have become famous for various reasons. In particular, certain generalizations of the polyhedral groups, scattered among the numerous papers of G. A. MILLER, are derived as members of a single family. The inclusion of a slightly different generalization in § 6.7 is justified by its unexpected connection with SHEPHARD's regular complex polygons.

Chapter 8 pursues BRAHANA's idea that any group generated by two elements, one of period 2, can be represented by a regular map or topological polyhedron.

In Chapter 9 we prove that every finite group defined by relations of the form

$$R_i^2 = (R_i R_j)^{p_{ij}} = E \qquad (1 \leqq i < j \leqq n)$$

can be represented in Euclidean n-space as a group generated by reflections in n hyperplanes. Many well-known groups belong to this family. Some of them play an essential role in the theory of simple Lie groups.

We wish to express our gratitude to Professor REINHOLD BAER for inviting us to undertake this work and for constructively criticizing certain parts of the manuscript. In the latter capacity we would extend our thanks also to Dr. PATRICK DU VAL, Professor IRVING REINER, Professor G. DE B. ROBINSON, Dr. F. A. SHERK, Dr. J. A. TODD and Professor A. W. TUCKER. We thank Mr. J. F. PETRIE for two of the drawings: Figs. 4.2, 4.3; and we gratefully acknowledge the assistance of Mrs. BERYL MOSER in preparing the typescript.

University of Toronto H. S. M. C.
University of Saskatchewan W. O. J. M.

February 1957

Preface to the Second Edition

We are grateful to Springer-Verlag for undertaking the publication of a revised edition, and to the many readers of the first edition who made suggestions for improvement. We have added to § 2.2 a brief account of the use of electronic computers for enumerating cosets in a finite abstract group. In § 6.5, the binary polyhedral groups are now more fully described. In § 6.8, recent progress on the Burnside problem has been recorded. New presentations for $GL(2, p)$ and $PGL(2, p)$ (for an odd prime p) have been inserted in § 7.5. In § 7.8, the number of relations needed for the Mathieu group M_{11} is reduced from 8 to 6; a presentation is now given also for M_{12}. Several new regular maps have been added to Chapter 8. There are also some improvements in § 9.7 and Table 2, as well as numerous small corrections.

University of Toronto H. S. M. C.
McGILL University W. O. J. M.

September 1964

Preface to the Third Edition

Although many pages of the Second Edition have been reproduced without alteration, there are about eighty small improvements in addition to the following. The section on BURNSIDE's problem (§ 6.8) now includes LEECH's presentation for $B_{3,3}$ and the important results of ADJAN and NOVIKOV on $B_{m,n}$ for large values of n. The section on $LF(2, p)$ (§ 7.5) has been almost entirely re-written because the number of relations needed to define this group no longer increases with p; the new presentations are surprisingly concise. The section on the MATHIEU groups has been improved in a similar manner.

Until recently, the deduction of 6.521 from 6.52 (page 68) had been achieved only by separate consideration of the separate cases. A general treatment, along the lines of Chapter 3, has been given by J. H. CONWAY, H. S. M. COXETER and G. C. SHEPHARD in Tensor **25** (1972), 405—418. An adequate summary of this work would have unduly increased the length of our book. For the same reason we have scarcely mentioned the important book by MAGNUS, KARRASS and SOLITAR (1966).

January 1972 H. S. M. C. W. O. J. M.

Preface to the Fourth Edition

Apart from many small corrections, the principal change from the Third Edition is a revised Chapter 2. The process of coset enumeration is now explained more clearly, and is applied to the problem of finding a presentation for a subgroup. To avoid lengthening the chapter, we have transferred four worked examples to the Appendix on pages 143—148.

Another innovation (at the end of page 79) is J. G. SUNDAY's combinatorial interpretation for the number q in the symbol $l\{q\}m$ for a regular complex polygon. Table 5 (on page 137) now includes a surprisingly neat presentation for the alternating group of degree 7.

University of Toronto H. S. M. C.

McGILL University W. O. J. M.

May 1979

Contents

Chapter 1

Cyclic, Dicyclic and Metacyclic Groups

After briefly defining such fundamental concepts as generators, factor groups and direct products, we show how an automorphism of a given group enables us to adjoin a new element so as to obtain a larger group; e.g., the cyclic and non-cyclic groups of order 4 yield the quaternion group and the tetrahedral group, respectively. Observing that the standard treatises use the term *metacyclic* group in two distinct senses, we exhibit both kinds among the groups of order less than 32, whose simplest known abstract definitions are collected in Table 1.

Opinions seem to be evenly divided as to whether products of group elements should be read from left to right or from right to left. We choose the former convention, so that, if A and B are transformations, AB signifies the transformation A followed by B.

1.1 Generators and relations. Certain elements $S_1, S_2, \ldots, S_m$, of a given discrete group $\mathfrak{G}$, are called a set of *generators* if every element of $\mathfrak{G}$ is expressible as a finite product of their powers (including negative powers). Such a group is conveniently denoted by the symbol

$$\{S_1, S_2, \ldots, S_m\}.$$

When $m = 1$, we have a *cyclic* group

$$\{S\} \cong \mathfrak{C}_q,$$

whose order q is the period of the single generator S. If q is finite, S satisfies the relation $S^q = E$, where E denotes the identity element.

A set of relations

$$g_k(S_1, S_2, \ldots, S_m) = E \qquad (k = 1, 2, \ldots, s), \tag{1.11}$$

satisfied by the generators of $\mathfrak{G}$, is called an *abstract definition* or *presentation* of $\mathfrak{G}$ if every relation satisfied by the generators is an algebraic consequence of these particular relations. For instance, if q is finite, $S^q = E$ is an abstract definition of $\mathfrak{C}_q$. It is important to remember that, in such a context, the relation $S^q = E$ means that the period of S is exactly q, and not merely a divisor of q. This is sometimes expressed by saying that the relation is not merely "satisfied" but "fulfilled" (see Miller, Blichfeldt and Dickson 1916, p. 143).

Returning to the general group $\mathfrak{G}$, defined by 1.11, let $\mathfrak{H}$ be another group whose abstract definition in terms of generators $T_1, T_2, \ldots, T_n$ is given by the relations

$$h_l(T_1, T_2, \ldots, T_n) = E \qquad (l = 1, 2, \ldots, t). \tag{1.12}$$

Then it is known that a necessary and sufficient condition for $\mathfrak{G}$ to be isomorphic to $\mathfrak{H}$ is the existence of relations

$$T_j = T_j(S_1, S_2, \ldots, S_m) \qquad (j = 1, 2, \ldots, n), \tag{1.13}$$

$$S_i = S_i(T_1, T_2, \ldots, T_n) \qquad (i = 1, 2, \ldots, m), \tag{1.14}$$

such that 1.11 and 1.13 together are algebraically equivalent to 1.12 and 1.14 together (COXETER 1934b). For instance,

$$R^6 = E \tag{1.15}$$

and

$$S^3 = T^2 = S^{-1}TST = E \tag{1.16}$$

are two possible presentations for $\mathfrak{C}_6$, since the relations

$$R^6 = E,\ S = R^4,\ T = R^3$$

are equivalent to

$$S^3 = T^2 = S^{-1}TST = E,\ R = ST.$$

1.2 Factor groups. Let $\mathfrak{G}' \cong \{R_1, R_2, \ldots, R_m\}$ be defined by the $s + r$ relations

$$g_k(R_1, R_2, \ldots, R_m) = E \qquad (k = 1, 2, \ldots, s + r).$$

The correspondence

$$S_i \to R_i \qquad (i = 1, 2, \ldots, m)$$

defines a homomorphism of $\mathfrak{G}$ (defined by 1.11) onto $\mathfrak{G}'$. The elements

$$g_k(S_1, S_2, \ldots, S_m) \qquad (k = s + 1, \ldots, s + r) \tag{1.21}$$

of $\mathfrak{G}$ all correspond to the identity element

$$g_k(R_1, R_2, \ldots, R_m) = E \qquad (k = s + 1, \ldots, s + r)$$

of $\mathfrak{G}'$. Hence the kernel of the homomorphism is the normal subgroup

$$\mathfrak{N} \cong \{W^{-1} g_k(S_1, S_2, \ldots, S_m)\, W\} \qquad (k = s + 1, \ldots, s + r),$$

where W runs through all the elements of $\mathfrak{G}$. In fact, $\mathfrak{N}$ is the smallest normal subgroup of $\mathfrak{G}$ that contains the elements 1.21, and it follows that

$$\mathfrak{G}' \cong \mathfrak{G}/\mathfrak{N}.$$

In other words, the effect of adding new relations to the abstract definition of a group $\mathfrak{G}$, is to form a new group $\mathfrak{G}'$ which is a *factor group* of $\mathfrak{G}$.

In particular, the effect of adding to 1.11 the relations

$$S_i^{-1}S_j^{-1}S_iS_j = E \qquad (i, j = 1, 2, \ldots, m)$$

is to form the *commutator quotient group* of $\mathfrak{G}$, which is the largest Abelian factor group of $\mathfrak{G}$.

Every group with m generators is a factor group of the free group $\mathfrak{F}_m$, which has m generators and no relations (REIDEMEISTER 1932a, p. 31). Apart from some special considerations in § 7.3, p. 88, we shall not attempt to describe the modern development of the theory of free groups, which began with the remarkable theorem of NIELSEN (1921) and SCHREIER (1927) to the effect that *every subgroup of a free group is free* (see especially MAGNUS 1939; BAER 1945; CHEN 1951, 1954; FOX 1953, 1954; KUROSCH 1953, pp. 271—274; M. HALL 1959, p. 96).

1.3 Direct products. If two groups $\mathfrak{G}$, $\mathfrak{H}$, defined by the respective sets of relations 1.11, 1.12, have no common element except E, and if all elements of $\mathfrak{G}$ commute with those of $\mathfrak{H}$, then the $m + n$ elements S_i and T_j generate the *direct product*

$$\mathfrak{G} \times \mathfrak{H}.$$

Clearly, a sufficient abstract definition is provided by 1.11, 1.12, and

$$S_i^{-1}T_j^{-1}S_iT_j = E \quad (i = 1, \ldots, m; j = 1, \ldots, n).$$

However, in many cases the number of generators may be reduced and the relations simplified. As an example, consider the cyclic groups $\mathfrak{C}_3$ and $\mathfrak{C}_2$, defined by the respective relations

$$S^3 = E \quad \text{and} \quad T^2 = E.$$

Their direct product $\mathfrak{C}_3 \times \mathfrak{C}_2$, of order 6, has the abstract definition 1.16; but it is also generated by the single element $R = ST$ and defined by the single relation 1.15, which shows that $\mathfrak{C}_3 \times \mathfrak{C}_2 \cong \mathfrak{C}_6$. More generally, the direct product of cyclic groups of orders q and r is an Abelian group $\mathfrak{C}_q \times \mathfrak{C}_r$, of order qr, which is cyclic if q and r are coprime:

$$\mathfrak{C}_q \times \mathfrak{C}_r \cong \mathfrak{C}_{qr}, \qquad (q, r) = 1.$$

Still more generally, if $p, q, \ldots$ are distinct primes, any Abelian group of order

$$p^\alpha q^\beta \ldots$$

is a direct product

$$\mathfrak{G}_{p^\alpha} \times \mathfrak{G}_{q^\beta} \times \cdots$$

of Abelian p-groups (BURNSIDE 1911, pp. 100—107), and every such p-group is a direct product of cyclic groups:

$$\mathfrak{G}_{p^\alpha} \cong \mathfrak{C}_{p^{\alpha_1}} \times \mathfrak{C}_{p^{\alpha_2}} \times \cdots,$$

where

$$\alpha = \alpha_1 + \alpha_2 + \cdots.$$

This p-group is described as the Abelian group of order p^α and type $(\alpha_1, \alpha_2, \ldots)$; in particular, the direct product of α cyclic groups of order p is the Abelian group of order p^α and type $(1, 1, \ldots, 1)$:

$$\mathfrak{C}_p^\alpha \cong \mathfrak{C}_p \times \mathfrak{C}_p \times \cdots \times \mathfrak{C}_p.$$

Combining the above results, we see that every finite Abelian group is a direct product of cyclic groups.

The infinite cyclic group $\mathfrak{C}_\infty$ is generated by a single element X without any relations. Thus it is the same as the free group $\mathfrak{F}_1$ on one generator. The inverse X^{-1} is the only other element that will serve as a generator. The direct product

$$\mathfrak{C}_\infty^2 \cong \mathfrak{C}_\infty \times \mathfrak{C}_\infty$$

of two infinite cyclic groups is defined by the single relation

$$XY = YX. \tag{1.31}$$

Its finite factor groups are obtained by adding relations of the type

$$X^b Y^c = E.$$

For example, in the Abelian group

$$X^b Y^c = X^{-c} Y^b = E, \; XY = YX, \tag{1.32}$$

we have

$$X^c = Y^b, \; X^{c^2} = Y^{bc} = X^{-b^2},$$

and therefore $X^n = E$, where $n = b^2 + c^2$. Suppose $(b, c) = d = \gamma b - \beta c$. Then X^d is a power of Y, namely

$$X^d = X^{\gamma b - \beta c} = Y^{-(\beta b + \gamma c)}.$$

Also

$$Y^b = X^c = Y^{-c(\beta b + \gamma c)/d} = Y^{b - \gamma n/d},$$

$$Y^c = X^{-b} = Y^{b(\beta b + \gamma c)/d} = Y^{c + \beta n/d}.$$

Since $(\beta, \gamma) = 1$, the period of Y divides n/d, and any element of the group is expressible as

$$X^x Y^y \qquad (0 \leq x < d, \; 0 \leq y < n/d).$$

Consider the direct product $\mathfrak{C}_d \times \mathfrak{C}_{n/d}$ in the form

$$Z^d = Y^{n/d} = E, \; ZY = YZ.$$

The element $X = Z Y^{-(\beta b + \gamma c)/d}$ satisfies $XY = YX$ and

$$X^b Y^c = Z^b Y^{\{-b(\beta b + \gamma c) + c(\gamma b - \beta c)\}/d} = Z^b Y^{-\beta n/d} = E,$$

$$X^{-c} Y^b = Z^{-c} Y^{\{c(\beta b + \gamma c) + b(\gamma b - \beta c)\}/d} = Z^{-c} Y^{\gamma n/d} = E.$$

Hence the group $\{X, Y\}$ of 1.32 is $\mathfrak{C}_d \times \mathfrak{C}_{n/d}$, the direct product of cyclic groups generated by

$$XY^{(\beta b + \gamma c)/d} \text{ and } Y.$$

It can be shown similarly that the Abelian group

$$X^c = Y^b = X^{-b}Y^{-c}, \; XY = YX$$

or

$$Y^c = Z^b, \; Z^c = X^b, \; XYZ = ZYX = E \qquad (1.33)$$

is $\mathfrak{C}_d \times \mathfrak{C}_{t/d}$, the direct product of cyclic groups generated by

$$XY^{(\beta b + \gamma c)/d} \text{ and } Y,$$

where $t = b^2 + bc + c^2$ and $d = (b, c) = \gamma b - \beta c$ (FRUCHT 1955, p. 12).

1.4 Automorphisms. Consider again the group $\mathfrak{G} \cong \{S_1, S_2, \ldots, S_m\}$ defined by 1.11. Suppose it contains m elements $S'_1, S'_2, \ldots, S'_m$ which satisfy the same relations

$$g_k(S'_1, S'_2, \ldots, S'_m) = E \qquad (k = 1, 2, \ldots, s)$$

but do not satisfy any further relations not deducible from these. Then the correspondence

$$S_i \to S'_i \qquad (i = 1, 2, \ldots, m) \qquad (1.41)$$

defines an *automorphism* of $\mathfrak{G}$.

One fruitful method for deriving a larger group $\mathfrak{G}^*$ from a given group $\mathfrak{G}$ is to adjoin a new element T, of period ac (say), which transforms the elements of $\mathfrak{G}$ according to an automorphism of period c. If we identify T^c with an element U of period a in the centre of $\mathfrak{G}$, left fixed by the automorphism, the order of $\mathfrak{G}^*$ is evidently c times that of $\mathfrak{G}$. If the automorphism is given by 1.41, the larger group is defined by the relations 1.11 and

$$T^{-1}S_iT = S'_i, \; T^c = U. \qquad (1.42)$$

This procedure is easily adapted to infinite groups. Although a may be infinite, $\mathfrak{G}$ is still a normal subgroup of index c in $\mathfrak{G}^*$.

In the case of an *inner* automorphism, $\mathfrak{G}$ contains an element R such that, for every S in $\mathfrak{G}$,

$$R^{-1}SR = T^{-1}ST,$$

i.e., $TR^{-1}S = STR^{-1}$. Thus the element

$$Z = TR^{-1} = R^{-1}T$$

of $\mathfrak{G}^*$ commutes with every element of $\mathfrak{G}$. The lowest power of Z that belongs to $\mathfrak{G}$ is

$$V = Z^c = UR^{-c},$$

of period b, say. This element V, like Z, commutes with every element of $\mathfrak{G}$; since it belongs to $\mathfrak{G}$, it belongs to the centre.

If $(b, c) = 1$ (for instance, if b is prime to the order of the centre, as in COXETER 1939, p. 90), consider integers β, γ, such that

$$\gamma b - \beta c = 1.$$

Instead of adjoining T to $\mathfrak{G}$, we could just as well adjoin $Z = TR^{-1}$, or adjoin

$$Z V^{\beta} = Z^{1+\beta c} = Z^{\gamma b},$$

whose cth power is

$$Z^{\gamma b c} = V^{\gamma b} = E$$

(since $V^b = E$). Hence in this case

$$\mathfrak{G}^* \cong \mathfrak{G} \times \mathfrak{C}_c, \tag{1.43}$$

where $\mathfrak{C}_c$ is the cyclic group generated by $Z^{\gamma b}$.

1.5 Some well-known finite groups. The cyclic group $\mathfrak{C}_q$, defined by the single relation

$$S^q = E, \tag{1.51}$$

admits an outer automorphism of period 2 which transforms every element into its inverse. Adjoining a new element R_1, of the same period, which transforms $\mathfrak{C}_q$ according to this automorphism, we obtain the *dihedral* group $\mathfrak{D}_q$, of order $2q$, defined by 1.51 and

$$R_1^{-1} S R_1 = S^{-1}, \; R_1^2 = E,$$

that is,

$$S^q = R_1^2 = (S R_1)^2 = E. \tag{1.52}$$

The same group $\mathfrak{D}_q$ is equally well generated by the elements R_1 and $R_2 = R_1 S$, in terms of which its abstract definition is

$$R_1^2 = R_2^2 = (R_1 R_2)^q = E. \tag{1.53}$$

The "even" dihedral group $\mathfrak{D}_{2m}$, defined by 1.53 with $q = 2m$, has a centre of order 2 generated by $Z = (R_1 R_2)^m$. If m is odd, the two elements R_1 and $R = R_2 Z$ satisfy the relations

$$R_1^2 = R^2 = (R_1 R)^m = E,$$

so that $\{R_1, R\}$ is $\mathfrak{D}_m$, and we have

$$\mathfrak{D}_{2m} \cong \mathfrak{C}_2 \times \mathfrak{D}_m \qquad (m \text{ odd}). \tag{1.54}$$

Since $\mathfrak{D}_{2m}$ (m odd) can be derived from $\mathfrak{D}_m$ by adjoining R_2, which transforms $\mathfrak{D}_m$ in the same manner as R, we see that 1.54 is an example of 1.43 (with $a = b = 1$, $c = 2$, $\beta = \gamma = -1$, $T = R_2$, $U = V = E$).

When $m = 1$, 1.54 is the *four-group*

$$\mathfrak{D}_2 \cong \mathfrak{C}_2 \times \mathfrak{D}_1 \cong \mathfrak{C}_2 \times \mathfrak{C}_2,$$

defined by
$$R_1^2 = R_2^2 = (R_1 R_2)^2 = E.$$

In terms of the three generators R_1, R_2 and $R_0 = R_1R_2$, these relations become
$$R_0^2 = R_1^2 = R_2^2 = R_0 R_1 R_2 = E. \qquad (1.55)$$

In this form, $\mathfrak{D}_2$ clearly admits an outer automorphism of period 3 which cyclically permutes the three generators. Adjoining a new element S which transforms $\mathfrak{D}_2$ in this manner, we obtain a group of order 12 defined by 1.55 and
$$S^3 = E, \; S^{-i} R_0 S^i = R_i \qquad (i = 1, 2).$$

The same group is generated by S and R_0, in terms of which it has the abstract definition
$$S^3 = R_0^2 = (S R_0)^3 = E. \qquad (1.56)$$

Since the permutations $S = (1\,2\,3)$ and $R_0 = (1\,2)\,(3\,4)$ generate the *alternating* group $\mathfrak{A}_4$ of order 12 and satisfy the relations 1.56, we conclude that the group defined by these relations is $\mathfrak{A}_4$. The above derivation shows that $\mathfrak{A}_4$ contains $\mathfrak{D}_2$ as a normal subgroup.

$\mathfrak{A}_4$ is equally well generated by S and $U = S^{-1}R_0$, in terms of which its definition is
$$S^3 = U^3 = (S U)^2 = E. \qquad (1.57)$$

Clearly, $\mathfrak{A}_4$ admits an outer automorphism of period 2 which interchanges the generators S and U. Adjoining such an element T, we obtain a group of order 24 defined by 1.57 and
$$T^2 = E, \; TST = U. \qquad (1.58)$$

In terms of the generators S and T, this group is defined by
$$S^3 = T^2 = (S T)^4 = E. \qquad (1.59)$$

Since the permutations $S = (2\,3\,4)$, $T = (1\,2)$ generate the *symmetric* group $\mathfrak{S}_4$ of order 24 and satisfy 1.59, we conclude that the group defined by these relations is $\mathfrak{S}_4$. In terms of the generators S and $U = S^{-1} T$, $\mathfrak{S}_4$ is defined by
$$S^3 = U^4 = (S U)^2 = E.$$

1.6 Dicyclic groups. When q is even, say $q = 2m$, the automorphism $S \to S^{-1}$ of $\mathfrak{C}_q$ can be used another way. Adjoining to
$$S^{2m} = E \qquad (1.61)$$
a new element T, of period 4, which transforms S into S^{-1} while its square is S^m, we obtain the *dicyclic* group $\langle 2, 2, m \rangle$, of order $4m$, defined by 1.61 and
$$T^2 = S^m, \; T^{-1} S T = S^{-1}.$$

Since the last relation may be written as $(ST)^2 = T^2$, S and T satisfy

$$S^m = T^2 = (ST)^2. \tag{1.62}$$

To show that these two relations suffice to define $\langle 2, 2, m\rangle$, we observe that they imply

$$S^m = T^2 = T^{-1}T^2T = T^{-1}S^mT = (T^{-1}ST)^m = S^{-m},$$

which is 1.61 (COXETER 1940c, p. 372; cf. MILLER, BLICHFELDT and DICKSON 1916, p. 62).

In terms of the three generators S, T, and $R = ST$, $\langle 2, 2, m\rangle$ is defined by the relations

$$R^2 = S^m = T^2 = RST, \tag{1.63}$$

or in terms of R and T alone:

$$R^2 = T^2 = (R^{-1}T)^m. \tag{1.64}$$

Of course, the symbol $\langle 2, 2, m\rangle$ could just as well have been written as $\langle m, 2, 2\rangle$ or $\langle 2, m, 2\rangle$. Other groups $\langle l, m, n\rangle$ will be discussed in § 6.5.

1.7 The quaternion group. The smallest dicyclic group

$$\mathfrak{Q} \cong \langle 2, 2, 2\rangle,$$

called the *quaternion* group, is defined by $S^2 = T^2 = (ST)^2$ or

$$R^2 = S^2 = T^2 = RST. \tag{1.71}$$

Note the resemblance to the famous formula

$$i^2 = j^2 = k^2 = ijk = -1$$

of HAMILTON (1856, p. 446).

$\mathfrak{Q}$ is the smallest *Hamiltonian* group, that is, it is the smallest non-Abelian group all of whose subgroups are normal. In fact, the finite Hamiltonian groups are precisely the groups of the form

$$\mathfrak{Q}\times\mathfrak{A}\times\mathfrak{B},$$

where $\mathfrak{A}$ is an Abelian group of odd order, and $\mathfrak{B}$ is an Abelian group of order 2^m ($m \geqq 0$) and type $(1, 1, \ldots, 1)$ (DEDEKIND 1897; HILTON 1908, p. 177; CARMICHAEL 1937, p. 114; ZASSENHAUS 1958, p. 160; SCORZA 1942, p. 89).

$\mathfrak{Q}$ is also the smallest group of ***rank*** 1, that is, it is the smallest non-Abelian group all of whose proper subgroups are Abelian. The groups of rank 1 have been investigated by MILLER and MORENO (1903), SCHMIDT (1924) and RÉDEI (1947). RÉDEI showed that, apart from $\mathfrak{Q}$, every such group belongs to one of three well-defined families. It thus appears that $\mathfrak{Q}$ is the only finite non-Abelian group all of whose proper subgroups are Abelian and normal.

1.8 Cyclic extensions of cyclic groups. If $(q, r) = 1$, the cyclic group 1.51 admits an automorphism

$$S \to S^r, \tag{1.81}$$

whose period c is the exponent to which r belongs modulo q, so that

$$r^c \equiv 1 \pmod{q}.$$

We derive a group of order qc by adjoining a new element T, of period ac (where a divides both q and $r-1$), such that

$$T^{-1}ST = S^r,\ T^c = S^{q/a}.$$

Writing $m = q/a$, we have the abstract definition

$$S^m = T^c = U,\ U^a = E,\ T^{-1}ST = S^r \tag{1.82}$$

(implying $U^r = S^{rm} = (T^{-1}ST)^m = T^{-1}UT = U$) for this group of order mac. (When $c = 2$ and $r = -1$, the group is dihedral or dicyclic according as $a = 1$ or $a = 2$.)

These relations can be simplified if

$$(a, m) = 1.$$

For if $\mu a + \alpha m = 1$, we have

$$S = S^{\mu a + \alpha m} = S^{\mu a} U^{\alpha} = (S^a)^{\mu}\, T^{c\alpha},$$

so that 1.82 is generated by $S_1 = S^a$ and T, in terms of which it has the abstract definition

$$S_1^m = T^{ac} = E,\ T^{-1}S_1T = S_1^r.$$

Dropping the subscript, we are thus led to consider the group

$$S^m = T^n = E,\ T^{-1}ST = S^r, \tag{1.83}$$

of order mn, derived from $\mathfrak{C}_m$ by adjoining T, of period n, which transforms $\mathfrak{C}_m$ according to the automorphism 1.81, of period c. The new feature is that we no longer identify T^c with an element of $\mathfrak{C}_m$. Since

$$S^{r^n} = T^{-n}ST^n = S,$$

the consistency of the relations 1.83 requires

$$r^n \equiv 1 \pmod{m}; \tag{1.84}$$

i.e., n must be a multiple of the exponent to which r belongs modulo m (CARMICHAEL 1937, p. 176). Thus 1.83 is a factor group of

$$T^n = E,\ T^{-1}ST = S^r \tag{1.85}$$

(where r may be positive or negative). The group 1.85 is infinite if $r^n = 1$, and of order

$$n \cdot |r^n - 1|$$

otherwise. When $r = 1$, 1.83 is obviously $\mathfrak{C}_m \times \mathfrak{C}_n$. When $r = -1$ and n is odd, 1.84 implies $m = 2$, so that 1.85 is $\mathfrak{C}_2 \times \mathfrak{C}_n \cong \mathfrak{C}_{2n}$.

The centre of 1.83 evidently contains T^c, where c is the exponent to which r belongs modulo m. If

$$(a, c) = 1,$$

where $a = n/c$, we can express T in terms of $T_1 = T^a$ and T^c, so that 1.83 is the direct product of

$$S^m = T_1^c = E, \quad T_1^{-1} S T_1 = S^{r^a}$$

with the $\mathfrak{C}_a$ generated by T^c. For instance, if a is odd, the group

$$S^m = T^{2a} = E, \quad T^{-1} S T = S^{-1} \qquad (1.86)$$

is the direct product

$$\mathfrak{D}_m \times \mathfrak{C}_a.$$

(On the other hand, when m is odd and $a = 2$, 1.86 is $\langle 2, 2, m \rangle$.)

The period of S, being a divisor of $r^n - 1$, may or may not be a multiple of $r - 1$. For a discussion of the former case, it is convenient to make a slight change of notation. In the group

$$R^{(r-1)m} = T^n = E, \quad T^{-1} R T = R^r, \qquad (1.861)$$

T transforms R^m into $R^{rm} = R^m$; thus R^m belongs to the centre. If

$$(r - 1, m) = 1, \qquad (1.87)$$

we can express R in terms of $S = R^{r-1}$ and R^m, so that in this case $\{R, T\}$ is the direct product of 1.83 with the $\mathfrak{C}_{r-1}$ generated by R^m. For instance, when m is odd, the group

$$R^{2m} = T^4 = E, \quad T^{-1} R T = R^{-1}$$

is $\mathfrak{C}_2 \times \langle 2, 2, m \rangle$.

We are thus led to consider the group 1.83 under the special condition 1.87. Since S is now a power of the commutator

$$S^{-1} T^{-1} S T = S^{-1} S^r = S^{r-1},$$

the commutator subgroup is the $\mathfrak{C}_m$ generated by S. Its quotient group is defined by 1.83 and

$$ST = TS,$$

implying $S^{r-1} = E = S^m$, whence, by 1.87, $S = E$; thus the commutator quotient group is the $\mathfrak{C}_n$ defined by $T^n = E$.

A group whose commutator subgroup and commutator quotient group are cyclic is conveniently called *Z-metacyclic*, i.e., metacyclic in the sense of ZASSENHAUS (1958, p. 174). Thus we have proved that, if 1.84 and 1.87 are satisfied, 1.83 is Z-metacyclic. ZASSENHAUS proved that, conversely, *every finite Z-metacyclic group is expressible in this form.*

NEUMANN (1956, p. 191) has observed that this presentation of the general Z-metacyclic group is equivalent to the two relations

$$S^m = T^n, \quad T^{-1}S^uT = S^{u+1}$$

where u is the inverse of $r-1$ modulo m, so that

$$u(r-1) \equiv 1 \pmod{m}.$$

For instance, if m is odd, 1.86 is equivalent to

$$S^m = T^{2a}, \quad T^{-1}S^{(m-1)/2}T = S^{(m+1)/2}.$$

Let us call a Z-metacyclic group *ZS-metacyclic* if all its Sylow subgroups are cyclic. For this, ZASSENHAUS (1958, p. 174) found the necessary and sufficient condition

$$(m, n) = 1. \tag{1.88}$$

Thus the "odd" dihedral and "odd" dicyclic groups are ZS-metacyclic. But if $(m, a) > 1$, $\mathfrak{D}_m \times \mathfrak{C}_a$ (a odd) is only Z-metacyclic; e.g., since $\mathfrak{D}_3 \times \mathfrak{C}_3$ contains no element of period 9, its Sylow subgroup of order 9 is non-cyclic.

The most important case of a ZS-metacyclic group is

$$S^p = T^{p-1} = E, \quad T^{-1}ST = S^r, \tag{1.89}$$

where p is a prime and r is a primitive root (mod p). As NETTO (1900, p. 284) attributes it to KRONECKER, let us call this *K-metacyclic* (see also HEFFTER 1898; MILLER, BLICHFELDT and DICKSON 1916, p. 12; CARMICHAEL 1937, pp. 42, 184; M. HALL 1959, pp. 145—148).

The term *metacyclic* was used in yet another sense by WEBER (1895, p. 698).

1.9 Groups of order less than 32. We have seen that the group 1.83 is $\mathfrak{C}_m \times \mathfrak{C}_n$ when $r = 1$, and that the cases when $r = -1$ (or $m-1$) include:

$\mathfrak{C}_{2n}$	when n is odd (so that $m = 2$),
$\mathfrak{D}_m \times \mathfrak{C}_{n/2}$	when $n/2$ is odd (e.g., when $n = 2$),
$\langle 2, 2, m\rangle$	when m is odd and $n = 4$,
$\mathfrak{C}_2 \times \langle 2, 2, m/2\rangle$	when $m/2$ is odd and $n = 4$.

Moreover, the same group with T inverted is obtained by changing r into r^{-1} (mod m). The remaining solutions of 1.84 with $mn < 32$ may be listed as follows:

r	-1	± 2	2	-2	3	-3	4	-4	5	-5
n	4	4	3	3	2	2	2	2	2	2
m	4	5	7	9	8	8	15	15	12	12

In six of these cases $r^n - 1 = m$, so that the two relations 1.85 suffice. In two cases $(r-1, m) = (m, n) = 1$, so that the group is ZS-metacyclic. One of these two cases is the K-metacyclic group 1.89 with $p = 5$.

The rest of the entries in Table 1 are taken from various sources, such as CAYLEY (1889), BURNSIDE (1911, pp. 157—161), MILLER (1911), Miss BURNS (1915), MILLER, BLICHFELDT and DICKSON (1916, pp. 143—168), CARMICHAEL (1937, pp. 166—187), and COXETER (1939, pp. 81, 83, 143).

For tables giving the number of groups of each order less than 162 (except 128), the reader may consult MILLER (1930), SENIOR and LUNN (1934), and HALL and SENIOR (1964). For instance, the number of groups of order 160 (including Abelian groups) is 238. However, PHILIP HALL has informed us that MILLER was mistaken in giving the number of groups of order 64 as 294; the correct number is 267.

Chapter 2

Systematic Enumeration of Cosets

E. H. MOORE (1897) and many others have computed the index of a subgroup in an abstract group by systematically enumerating the cosets. TODD and COXETER (1936) converted this into a mechanical technique, a useful tool with a wide range of applications. In § 2.1 we describe this algorithm, in § 2.2 we use it to determine an abstract definition for a given finite group, and in § 2.3 we describe some additional computation, performed alongside the enumeration, which determines an abstract definition for the subgroup.

2.1 Coset enumeration. Let $\mathfrak{G}$ be the abstract group with generators $S_1, S_2, \ldots, S_m$ and relations

$$g_k(S_1, S_2, \ldots, S_m) = E \qquad (k = 1, 2, 3, \ldots, s)\,. \tag{2.11}$$

Choose a set of elements

$$T_i = f_i(S_1, S_2, \ldots, S_m) \qquad (i = 1, 2, 3, \ldots, n) \tag{2.12}$$

and let $\mathfrak{H} = \{T_1, T_2, \ldots, T_n\}$ be the subgroup generated by them. A *coset* of $\mathfrak{H}$ in $\mathfrak{G}$ is defined to be the set of slements $\mathfrak{H}R$ obtained by multiplying the elements of $\mathfrak{H}$ on the right by an element R of $\mathfrak{G}$. Two cosets of the same subgroup either coincide or have no elements in common, and every element of $\mathfrak{G}$ belongs to at least one such coset. The object of coset enumeration is to determine a complete set of distinct cosets; when the computation terminates we will have such a set of cosets, denoted by numerals 1, 2, 3, ...

The coset 1 is defined as the subgroup $\mathfrak{H}$. As the computation proceeds each of the cosets 2, 3, 4, ... will be defined as a product $\alpha S_i^{\pm 1}$ where α is a previously defined coset. These coset numbers will be used as entries in $s + n$ rectangular tables. Each f_i of the set of subgroup generators 2.12, and each g_k of the set of relations 2.11, written in expanded form, serves as the heading for a table; e. g., if $(S_1{}^3 S_2^{-2})^2 = E$ is one of 2.11, the corresponding "relation" table has the heading

$$S_1 S_1 S_1 S_2^{-1} S_2^{-1} S_1 S_1 S_1 S_2^{-1} S_2^{-1},$$

and if $T_2 = S_2 S_1^{-1} S_3^2$ is one of 2.12 the corresponding "subgroup generator" table has the heading

$$S_2 S_1^{-1} S_3 S_3.$$

If a heading has c factors, the table will have $c + 1$ columns, with each factor occurring between two adjacent columns. Coset numbers will be entered into the tables, always subject to the condition that coset numbers α and β occur side by side, β to the right of α, in a pair of columns headed by $S_i^{\pm 1}$, only if $\alpha S_i^{\pm 1} = \beta$ and $\beta S_i^{\mp 1} = \alpha$.

Each relation table has an unspecified number of rows with the first and last column identical; each subgroup generator table has one row with a 1 in the first and last column because of the initial information

$$1 T_j = 1 \qquad (j = 1, 2, \ldots, n).$$

Before proceeding we make sure that the number 1 appears in every relation table in every essentially different position. The coset 2 is defined by $2 = 1 S_i^{\pm 1}$ for some group generator S_i; this may be done by inserting the number 2 in some suitable unoccupied place which is preceded or followed by the number 1. We now (and at all times) fill up as much as possible of all the tables, adding enough rows to ensure that 2 appears in every essentially different place in every relation table. Then we define the next coset, $3 = 1 S_j^{\pm 1}$ or $3 = 2 S_j^{\pm 1}$, and continue filling in the tables while satisfying the "essentially different positions" condition.

Consider, for example, enumeration of the cosets of $\mathfrak{H}$ in $\mathfrak{G}$, where

$$\mathfrak{G} = \{A, B\}, \qquad A^3 = B^3 = (AB)^2 = E,$$

$$\mathfrak{H} = \{X, Y\}, \qquad X = AB, \quad Y = BA.$$

The tables, with $1X = 1$ and $1Y = 1$ inserted, are:

	A		B	
1				1

	B		A	
1				1

	A		A		A	
1						1

	B		B		B	
1						1

	A		B		A		B	
1								1
		1						

Note that we have placed the symbol 1 (meaning $\mathfrak{H}$) in every essentially different position in the relation tables. Define coset $2 = 1A$ and insert into the tables; the row of the "AB" generator table closes and rewards us with $2B = 1$. With all available information inserted the tables are

A B	B A	A A A	B B B	A B A B
1 2 1	1 1	1 2 1	1 2 1	1 2 1 2 1
				1
				2 2

Define $3 = 1B$; the "BA" generator table yields $3A = 1$ and then the A^3 relation table yields $2A = 3$. Now the tables close up:

A B	B A	A A A	B B B	A B A B
1 2 1	1 3 1	1 2 3 1	1 3 2 1	1 2 1 2 1
				3 1 3 1 3
				2 3 2 3 2

There are 3 distinct cosets of $\mathfrak{H}$ in $\mathfrak{G}$.

A definition of a coset involves information of the type $\alpha S_i^{\pm 1} = \beta$. Such information is also obtained when a row of a table becomes complete; e. g., if a row has just one empty space, say

$$\begin{array}{cccc} \ldots & S_k & S_i & S_j^{-1} & \ldots \\ \ldots\alpha & \gamma & & \beta\ldots \end{array}$$

and $\gamma S_i = \delta$ is known, we insert δ in the blank space and are rewarded with

$$\delta S_j^{-1} = \beta \quad \text{or} \quad \beta S_j = \delta .$$

This may be new information in the sense that coset βS_j has not hitherto been identified. On the other hand it may happen that coset βS_j had earlier been identified, as coset ε, say; that is, $\beta S_j = \varepsilon$. This means that the numbers δ and ε represent the same coset. If these numbers are equal, we may continue the computation. If they are not equal we have a "collapse" or "coincidence"; we replace the larger of the two integers by the smaller throughout and note any other coincidences which may result. When all these have been adjusted, we may preceed in the usual manner. When all the tables are complete, the process is at end: the number of distinct cosets of $\mathfrak{H}$ in $\mathfrak{G}$ has been determined.

Here is a case of complete collapse:

$$\mathfrak{G} = \{R,S\}, \quad RS^2 = S^3R, \quad SR^2 = R^3S ;$$

$$\mathfrak{H} = \{X\}, \qquad X = R .$$

The tables with $1X = 1$ inserted are:

	R	S	S	$=$	S	S	S	R
1	1			1				
		1						
			1				1	1
					1			
						1		

	S	R	R	$=$	R	R	R	S
1				1	1	1	1	
	1	1	1					1

It is not necessary to change the relations to the form $RS^2R^{-1}S^{-3} = SR^2S^{-1}R^{-3} = E$; as set up, each table will have the same first column and the same last column.

Define successively $1S = 2$, $2R = 3$, $2S = 4$, $3S = 5$. At this stage the tables are:

	R	S	S	$=$	S	S	S	R
1	1	2	4	1	2	4		4
		1	2				3	2
			1				1	1
					1	2	4	
			3			1	2	3
2	3	5		2	4			
3	2	4		3	5			
		3	5					5
					3	5		
						3	5	
4				4				
	4							
5				5				
	5							

	S	R	R	$=$	R	R	R	S
1	2	3	2	1	1	1	1	2
	1	1	1					1
2	4		5	2	3	2	3	5
	3	2	3					3
3	5		4	3	2	3	2	4
4				4				
		4						
					4			
						4		
							4	
5				5				
		5						
					5			
						5		
							5	

We now define $6 = 5R$ and find, upon inserting this into the tables, that a complete collapse takes place, i. e., the tables imply $1 = 2 = 3 = 4 = 5 = 6$. Since $1S = 2 = 1$, it follows that S is a power of R; with this information the relations for $\mathfrak{G}$ imply $R = S = E$.

It is convenient to store all information of the type $\alpha S_i^{\pm 1} = \beta$ (whether obtained by definition or when a row closes) in a "multiplication table" whose columns are headed by $S_1, S_1^{-1}, S_2, S_2^{-1}, \ldots, S_m, S_m^{-1}$ and rows are indexed by 1, 2, 3, ...Then the information $\alpha S_i^{\pm 1} = \beta$ is stored by placing a β in "box" $b(\alpha, S_i^{\pm 1})$ where row α meets column $S_i^{\pm 1}$, and placing an α in $b(\beta, S_i^{\mp 1})$.

If new cosets are defined so that the rows of the multiplication table fill up, the "essentially different positions" condition may be relaxed; when a new coset is defined, simply start and end (in every relation table) a new row with this new coset number; Example 1 in the Appendix illustrates this.

The reader may like to test his skill by enumerating the five cosets of $\{V_1, V_2\}$ in

$$V_1^3 = V_2^3 = V_3^3 = (V_2V_3)^2 = (V_3V_1)^2 = (V_1V_2)^2 = E$$

(CARMICHAEL 1923, p. 255). Further examples may be found in TODD and COXETER (1936) and COXETER (1939, 1956).

The systematic enumeration of cosets is sufficiently mechanical for the use of an electronic computer. In recent years a number of people have programmed the method for automatic execution. For example, JANE WATSON (see COXETER 1970, p. 45) has obtained, in a few seconds, results that formerly took many hours of monotonous labour. The work of HASELGROVE on EDSAC 1 at Cambridge in 1953 is believed to be the first attempt to carry out coset enumeration on a digital computer. This was followed by FELSCH (1961) and TROTTER (1964). More recent progress in coset enumeration is reported by LEECH (1970) and particularly CANNON, DIMINO, HAVAS and WATSON (1973), where further useful references may be found.

2.2 Finding a presentation for a finite group. Let us suppose that the relations 2.11 are proposed as an abstract definition for a given finite group $\mathfrak{G}^*$ of order g^*. In practice these equations are obtained by taking a set of generators of $\mathfrak{G}^*$ and observing some of the relations satisfied by them. It is natural to select relations of a simple form, such as those which express the periods of the generators or of simple combinations of them. The number of relations to select is a matter for experiment and the "method of cosets" is the way we test the success of the experiment. If the group $\mathfrak{G}$ defined by 2.11 is not itself isomorphic to $\mathfrak{G}^*$, it possesses a normal subgroup $\mathfrak{G}_1$ such that the factor group $\mathfrak{G}/\mathfrak{G}_1$ is isomorphic to $\mathfrak{G}^*$. In all cases the order of $\mathfrak{G}$ is at least g^*. If we can verify that the order of $\mathfrak{G}$ is at most g^*, it will follow that $\mathfrak{G}$ is isomorphic to $\mathfrak{G}^*$. To test whether this is or is not the case, we pick out a subset $T_1, T_2, \ldots, T_n$ of elements of $\mathfrak{G}$ such that the relations 2.11 imply the relations

$$h_l(T_1, T_2, \ldots, T_n) = E \qquad (2.21)$$

which are already known to be an abstract definition for a group $\mathfrak{H}'$ of order $h < g^*$. Conceivably the relations 2.11 imply other relations between the T's independent of 2.21, in which case the subgroup $\mathfrak{H}$ of $\mathfrak{G}$ generated by the T's will not be $\mathfrak{H}'$ but some factor group $\mathfrak{H}'/\mathfrak{H}_1$. In all cases the order of $\mathfrak{H}$ is at most h. We now embark on an enumeration of the cosets of $\mathfrak{H}$ in $\mathfrak{G}$. If the process terminates and we find that the number of cosets does not exceed g^*/h, our experiment has been successful: $\mathfrak{G}^*$ and $\mathfrak{G}$ are not merely homomorphic but isomorphic.

If the proposed relations 2.11 are insufficient to define $\mathfrak{G}$ the tables will fail to close up after g^*/h cosets have been introduced. Conversely, if the tables fail to close up when g^*/h cosets have been defined, then either the relations 2.11 are insuffcent to determine $\mathfrak{G}^*$, or else there are some undiscovered coincidences among the cosets already introduced. If the tables are "nearly" complete, it is probably the latter circumstance which has occurred, and the introduction of a few more cosets will demonstrate the coincidences. If on the other hand there are large gaps in the tables, it is a sign that 2.11 are probably insufficient to define $\mathfrak{G}^*$. Example 2 in the Appendix illustrates the use of coset enumeration described in this section.

2.3 Finding a presentation for a subgroup. We now describe some additional computation, performed alongside the enumeration of the cosets of $\mathfrak{H}$ (generated by the T's of 2.12) in $\mathfrak{G}$ (the abstract group 2.11) which yields a set of relations in the T's sufficent to define $\mathfrak{H}$ abstractly. For this purpose it will be convenient to set up the one-row subgroup generator tables with an extra column e. g., if $T_2 = S_2 S_1^{-1} S_3^2$ (or equivalently $T_2^{-1} S_2 S_1^{-1} S_3^2 = E$) is one of the definitions 2.12, the corresponding one-row table will be headed by

T_2^{-1}	S_2	S_1^{-1}	S_3	S_3
1				1

We choose coset representatives, $\bar{\alpha}$ denoting the representative of coset α, as follows: $\bar{1} = E$, and when coset β is defined, say $\beta = \alpha S_i$, choose $\bar{\beta} = \bar{\alpha} S_i$. This information is stored in the multiplication table (whose rows are now indexed by 1, 2, 3, . . .) by placing $E\beta$ in box $b(\bar{\alpha}, S_i)$ and placing $E\alpha$ in box $b(\bar{\beta}, S_i^{-1})$. Whenever a row of a table becomes complete we were previously rewarded with information of the type $\gamma S_j^{\pm 1} = \delta$; now, however, there is sufficient information in the multiplication table (and the initial information $\bar{1} T_j = T_j \bar{1}$, $\bar{1} T_j^{-1} = T_j^{-1} \bar{1}$) to deduce

$$\bar{\gamma} S_j^{\pm 1} = W \bar{\delta} \tag{2.31}$$

where W is a word in the T's. Example 3 in the appendix illustrates the details of this computation. We store 2.31 by placing $W\delta$ in $b(\bar{\gamma}, S_j^{\pm 1})$ if this box is empty. If this box is occupied, by $U\bar{\varepsilon}$, say (where U is a word in the T's) we conclude that

$$W\bar{\delta} = U\bar{\varepsilon}\,.$$

If $\bar{\varepsilon} = \bar{\delta}$ we deduce that

$$W = U\,, \tag{2.32}$$

a relation (possibly trivial) among the T's. If $\bar{\varepsilon} \neq \bar{\delta}$ we now know that ε and δ represent the same coset; in this case, assuming $\delta < \varepsilon$, we replace ε by δ and $\bar{\varepsilon}$ by $U^{-1} W \bar{\delta}$ throughout. This will usually introduce new coincidences which we process before continuing with the computation.

When the tables are complete, the relations 2.32 which we have captured along the way suffice to define $\mathfrak{H}$ abstractly in terms of the T's, as in Examples 3 and 4 in the Appendix.

Furthermore, the algorithm provides the algebraic steps by which these relations, and the information in the multiplication table, are deduced from relations 2.11. The completed multiplication table can be used to express any element of $\mathfrak{G}$, a word in the S's, as a product $V\bar{\varepsilon}$, where V is a word in the T's and $\bar{\varepsilon}$ is the appropriate coset representative. Particularly striking is the case of complete collapse: the algorithm provides the algebraic derivation of the relations for $\mathfrak{G}$, and expressions for the S's, in terms of the T's. Further examples may be found in CAMPBELL (1965, 1970). Similar techniques have been described and exploited by many others, e. g., MENDELSOHN (1970), BENSON and MENDELSOHN (1966), HAVAS (1974a, b), JOHNSON (1976, pp. 75–85), BEETHAM and CAMPBELL (1976) and LEECH (1971).

2.4 The corresponding permutations. The method yields a representation of $\mathfrak{G}$ as a transitive permutation group. Thus, in the first example in the Appendix,

$$B=(1\ 2\ 4\ 8\ 14\ 13\ 6\ 3)\ (5\ 11\ 15\ 10\ 16\ 12\ 7\ 9),$$
$$C=(1\ 3\ 7\ 11\ 16\ 13\ 14\ 8\ 15\ 12\ 5\ 2)\ (4\ 9\ 6\ 10),$$

and in the second example

$$S=(1\ 2\ 3)\ (4\ 5\ 6)\ (7\ 8\ 9)\ (10\ 11\ 12),$$
$$U=(2\ 3\ 4\ 7\ 5)\ (6\ 9\ 10\ 11\ 8).$$

However, the relation between $\mathfrak{G}$ and the permutation group may well be only a homomorphism. In fact, it is a simple isomorphism only in the following cases: when $\mathfrak{H} \cong \{E\}$, so that the representation is regular, and when neither $\mathfrak{H}$ nor any proper subgroup of $\mathfrak{H}$ is normal in $\mathfrak{G}$ (CARMICHAEL 1937, p. 157; see also BURNSIDE 1911, Chapters X to XII).

Chapter 3

Graphs, Maps and Cayley Diagrams

The chief purpose of this chapter is to describe CAYLEY's representation of a group with given generators by a topological 1-complex or graph, whose vertices represent the elements of the group while certain sets of edges are associated with the generators. CAYLEY (1878a, b) proposed the use of colours to distinguish the edges associated with different generators (see BURNSIDE 1911, pp. 423—427 and the frontispiece). Instead, for the sake of easier printing, we use lines drawn in various styles: ordinary, broken, dotted, etc. After suitably embedding the graph into a surface, we obtain a map from which a set of defining relations for the group may be read off.

The Cayley diagram was rediscovered by DEHN (1910, pp. 140—146). Accordingly, some authors (e.g., BAER and LEVI 1936, pp. 392, 393) call it the "DEHNsche Gruppenbild". But CAYLEY's priority is indisputable, as he described it in the year of DEHN's birth.

3.1 Graphs. We assume that the reader is familiar with the idea of a finite or infinite *graph*, whose *vertices* (or *nodes*) are joined in pairs by directed *edges* (or *branches*). A graph is said to be *connected* if every pair of its vertices can be joined by a *path* along a set of consecutively adjacent edges. The same path described backwards is called the *inverse* path. A *circuit* is a path whose first and last vertices coincide. A *trivial* circuit consists of a path followed by its own inverse. A *tree* is a connected graph whose circuits are all trivial; in a finite tree the number of vertices is one more than the number of edges (KÖNIG 1936, p. 51).

We regard a circuit as being essentially unchanged if it is combined with a trivial circuit; in particular, all trivial circuits are equivalent. With this understanding, any circuit may be regarded as including a certain "fixed" vertex E, and the *product* of two circuits is obtained by describing them in succession (REIDEMEISTER 1932a, p. 99).

Any graph that is not a tree has a certain minimal set of *fundamental* circuits (KÖNIG 1936, p. 61) such that every circuit of the graph is expressible as a finite product of their powers (including negative powers).

Consider a connected graph having N_0 vertices, N_1 edges, and μ fundamental circuits. By removing an edge that belongs to only one of these circuits[1]), we obtain a new graph having N_0 vertices, $N_1 - 1$ edges, and $\mu - 1$ fundamental circuits. Repeating the process, we obtain eventually a graph having N_0 vertices, $N_1 - \mu$ edges, and no fundamental circuits. Since this is a tree, $N_0 = N_1 - \mu + 1$, that is,

$$\mu = N_1 - N_0 + 1 \tag{3.11}$$

(KÖNIG 1936, p. 53; ORE 1962, p. 67).

[1]) Any edge that belongs (non-trivially) to at least one fundamental circuit will serve. For if it belongs to more than one, we can replace all but one by their respective products with that one, to obtain a new set of fundamental circuits in which the given edge belongs to only one.

3.2 Maps. We define a *map* to be the decomposition of an unbounded surface into N_2 non-overlapping simply-connected regions (called *faces*) by N_1 arcs (called *edges*) joining pairs of N_0 points (called *vertices*). It is understood that no two edges have a common interior point, and that every edge belongs to just two faces[1] (HILBERT and COHN-VOSSEN 1932, pp. 254—276; BALL 1974, pp. 232—237; LEFSCHETZ 1949, pp. 72—85). The *Euler-Poincaré characteristic*

$$\chi = N_0 - N_1 + N_2 \tag{3.21}$$

is a property of the surface, independent of the map; e.g., it is obviously the same for the *dual* map which has N_0 faces, N_1 edges and N_2 vertices. The possible values of χ are:

$$2, 0, -2, -4, \ldots \text{ for an orientable surface},$$

$$1, 0, -1, -2, \ldots \text{ for a non-orientable surface}.$$

Each non-orientable surface can be derived from the corresponding orientable surface (called its two-sheeted *covering surface*) by suitably identifying pairs of points; e.g., the projective plane (or elliptic plane, $\chi = 1$) can be derived from the sphere ($\chi = 2$) by identifying antipodes. Any map drawn on the former corresponds to another map on the latter, having twice as many elements of each kind (THRELFALL 1932a, pp. 41—42).

The vertices and edges of a map evidently form a graph. Conversely (LINDSAY 1959), any connected graph can be embedded into a surface so as to form a map having the same vertices and edges. The embedding is seldom unique, but there must be at least one for which N_2, and therefore χ, is maximal. When the maximal characteristic is 2, the graph is said to be *planar* (WHITNEY 1933) since it can be embedded into a sphere and so (say by stereographic projection) into a plane.

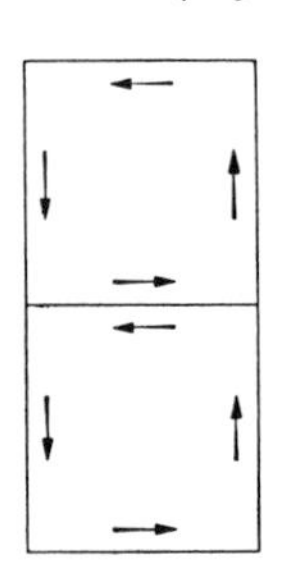

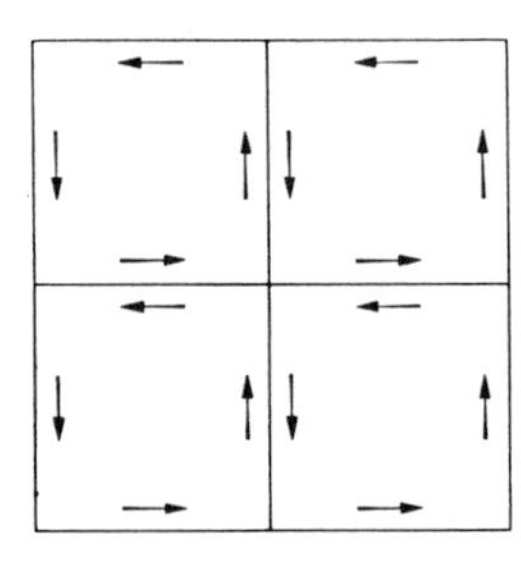

Fig. 3.2. Products of circuits

The circuit round a simply-connected region formed by two or more adjacent faces is evidently equal to the product of circuits (in the same sense) around the faces themselves (see Fig. 3.2). In particular, when a planar graph is embedded into a plane, one of the N_2 faces appears as a peri-

[1]) It may happen that the two faces coincide, i.e., that an edge occurs twice in the cycle of sides of one face. For instance, a tree forms a map on a sphere, having $N_1 + 1$ vertices, N_1 edges, and one face (a $2N_1$-gon).

pheral circuit which, surrounding the remaining $N_2 - 1$ faces, is equal to the product of circuits around them individually. In fact, these $N_2 - 1$ circuits are a fundamental system for the planar graph, so that

$$\mu = N_2 - 1.$$

Comparing this with 3.11, we verify EULER's formula

$$\chi = 2$$

for the sphere.

3.3 Cayley diagrams. CAYLEY (1878a, b) represented the multiplication table of a group $\mathfrak{G}$, with given generators, by a graph having one vertex for each element of the group. It is convenient to use the same symbol V for an element and its corresponding vertex. With each generator S_i, we associate a certain set of directed edges, say S_i-edges, the direction being indicated by an arrow. Two vertices, V and W, are joined by an S_i-edge, directed from V to W, whenever[1])

$$W = S_i V. \tag{3.31}$$

Thus at any vertex there are two edges for each generator, one directed towards the vertex and one directed away. The edges join the vertices to their "neighbours"; the neighbours of the vertex E represent the generators and their inverses.

Any path along the edges of the graph corresponds to a *word* (an element expressed as a product of the generators and their inverses); e.g., the path P along an S_2-edge forwards, then along an S_3-edge backwards, then along an S_1-edge forwards, corresponds to the word $P = S_1 S_3^{-1} S_2$. A path P leads from the vertex V to the vertex PV. If a word C satisfies the relation

$$C = E,$$

then any path C is a circuit, and conversely; e.g., if $S_1^n = E$, then any path $C = S_1^n$ is a cyclically directed n-gon.

Whenever a generator is *involutory* (i. e., of period two) the graph may be simplified by using a single undirected edge instead of the 2-circuit consisting of two oppositely directed edges joining the same two vertices.

The permutations of the vertices which preserve the naming of the incident edges are precisely those given by the right regular representation $\mathfrak{G}_r$ of the group $\mathfrak{G}$. For if such a permutation takes the vertices V and W of 3.31 into the vertices V' and W' respectively, then

$$W' = S_i V'.$$

---

[1]) Like BURNSIDE (1911, p. 423), we adopt the original convention of CAYLEY (1878b). But some authors would write $W = VS_i$ for which a precedent was provided by CAYLEY himself (1889).

Hence this permutation corresponds to the element

$$U = V^{-1}V' = W^{-1}W'$$

which takes the vertex E and its neighbours $S_i^{\pm 1}$ into the vertex U and its neighbours $S_i^{\pm 1}U$.

The following are examples of Cayley diagrams for some well known groups of low order.

For the alternating group $\mathfrak{A}_4$, generated by the permutations

$$S = (1\ 2\ 3),\ T = (1\ 2)\ (3\ 4),$$

the Cayley diagram can be drawn as in Fig. 3.3a (KEMPE 1886, p. 43).

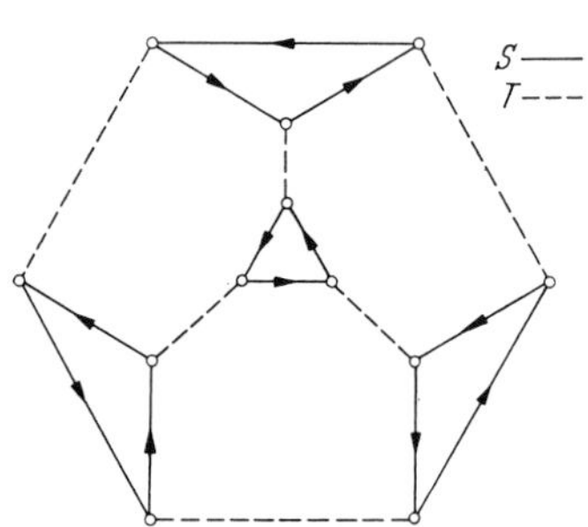

Fig. 3.3a. The tetrahedral group

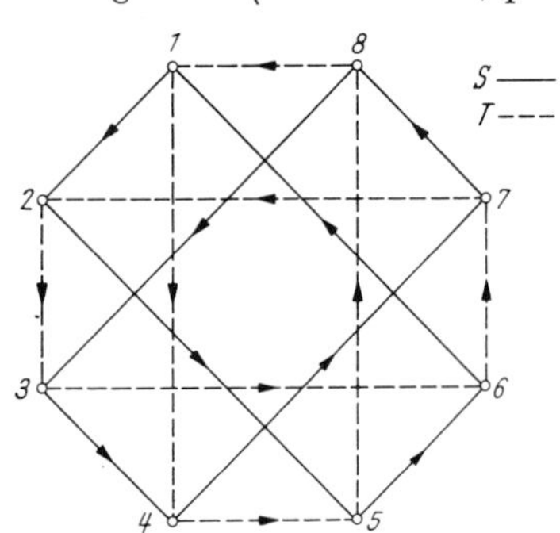

Fig. 3.3b. The quaternion group

Readers interested in the Archimedean solids will recognize this as a Schlegel diagram for the truncated tetrahedron (SCHLEGEL 1883, pp. 353—358).

For the quaternion group $\mathfrak{Q}$, in its regular representation

$$S = (1\ 2\ 5\ 6)\ (3\ 4\ 7\ 8),\ T = (2\ 3\ 6\ 7)\ (4\ 5\ 8\ 1),$$

the Cayley diagram can be drawn as an octagon 1 2 3 4 5 6 7 8 with an inscribed octagram 1 4 7 2 5 8 3 6 (Fig. 3.3b). For the group of order 16 generated by the permutations

$$R = (2\ 4)\ (3\ 6), S = (1\ 2)\ (3\ 5)\ (4\ 7)\ (6\ 8),$$

$$T = (1\ 3)\ (2\ 8)\ (4\ 5)\ (6\ 7),$$

we have an octagon joined to an octagram (Fig. 3.3c).

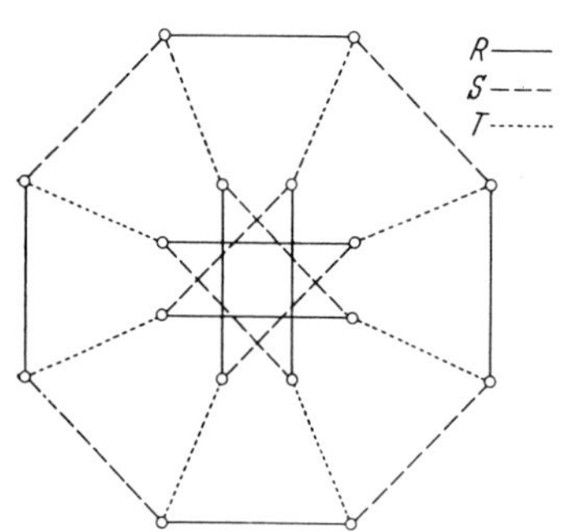

Fig. 3.3c. A group of order 16

For the group $\mathfrak{D}_3 \simeq \mathfrak{S}_3$, generated by the transpositions

$$R = (2\ 3),\ S = (3\ 1),\ T = (1\ 2),$$

the Cayley diagram is the Thomsen graph (Fig. 3.3d). The cyclic group $\mathfrak{C}_6$, generated by

$$R = (1\ 2\ 3\ 4\ 5\ 6),\ S = (1\ 4)\ (2\ 5)\ (3\ 6),$$

is represented by the same graph with six edges directed as in Fig. 3.3e.

In each of the last two cases the generator S is redundant, and by deleting all the S-edges we obtain a simpler diagram for the same group with fewer generators. In general, a generator S_i is redundant if and only if the deletion of all the S_i-edges leaves a connected graph. In particular, every Cayley diagram for a given finite group is a subgraph of the *complete* diagram (CAYLEY 1889) in which all the elements of the group

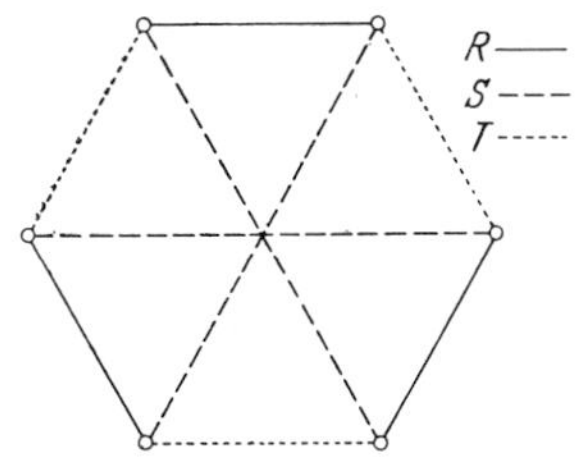

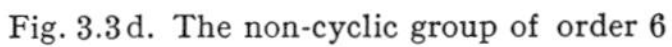
Fig. 3.3d. The non-cyclic group of order 6

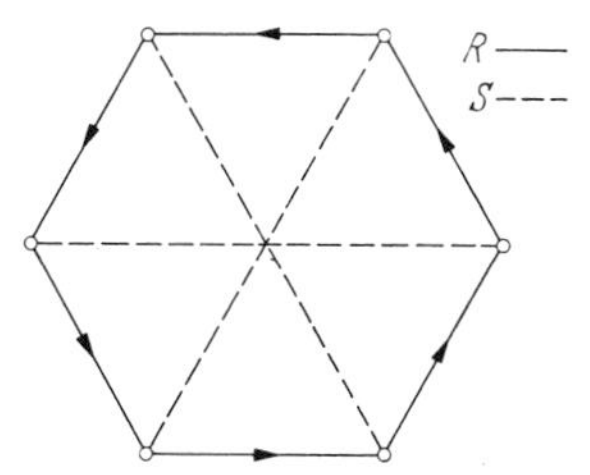

Fig. 3.3e. The cyclic group of order 6

(except E) are regarded as generators, and every vertex of the graph is joined twice to every other. Though too unwieldy to be useful in practice, this complete diagram is very easy to describe; any two distinct vertices, V and W, are joined by one edge directed from V to W and by another from W to V; the former is marked WV^{-1}, and the latter VW^{-1}.

3.4 Planar diagrams. We have seen that every circuit in a graph is expressible as a product of fundamental circuits (§ 3.1, p. 19), that the words in a group appear as paths in its Cayley diagram, that products of words appear as products of paths, and that a word is equal to E if and only if the corresponding path is a circuit. It follows that a sufficient (though usually redundant) set of defining relations is

$$C_1 = C_2 = \cdots = C_\mu = E, \tag{3.41}$$

where the C's are words corresponding to a set of fundamental circuits, including (because of our artificial simplification) the words S_i^2 corresponding to any undirected S_i-edge.

In particular, if the Cayley diagram is a planar graph, we can embed it in a plane and use the $N_2 - 1$ visible faces (along with any undirected edges).

A substantial reduction in the number of relations can be effected if the embedding happens to be *symmetrical*, i.e., if faces are taken into faces by the regular permutation group $\mathfrak{G}_r$. In this case, if V_1,

$V_2, \ldots, V_p$ are the vertices of a face, then for any element V, the vertices $V_1V, V_2V, \ldots, V_pV$ form a face. Since $\mathfrak{G}_r$ is transitive on the vertices, all of them are surrounded alike by faces; thus in writing out the relations 3.41, we may restrict attention to the faces at a single vertex.

The Cayley diagram shown in Fig. 3.3a is symmetrically embedded if we regard the outside region of the plane as one more (hexagonal) face, i.e., if we regard the figure not merely as a Schlegel diagram but as the surface of a solid truncated tetrahedron. The faces (and the undirected edge) at a vertex yield the relations

$$S^3 = T^2 = (ST)^3 = (TS)^3 = E,$$

one of which is obviously redundant. Thus $\mathfrak{A}_4$ has the abstract definition

$$S^3 = T^2 = (ST)^3 = E,$$

in agreement with 1.56.

When the Cayley diagram is symmetrically embedded into some surface other than a sphere (or plane), the faces at a vertex still yield true relations, but these generally do not suffice to define the given group. What further relations are needed? Before answering this question, let us summarize the topological theory of unbounded surfaces.

3.5 Unbounded surfaces. In Poincaré's theory, two directed circuits on a given surface are considered to be equivalent if they are *homotopic*, i.e., if one can be continuously transformed into the other (Lefschetz 1949, p. 159). The product of two circuits is obtained by distorting one of them (if necessary) until they have a common point, and then describing them in succession. In this sense, the classes of homotopic circuits are the elements of a group called the *fundamental group* of the surface (Kerékjártó 1923, pp. 11, 178; Hilbert and Cohn-Vossen 1932, p. 290). Its identity element is the class of circuits that bound simply-connected regions, i.e., of circuits that can be continuously shrunk to a point. Thus the fundamental group of a sphere is the group of order 1.

For a *torus*, or "sphere with a handle", the fundamental group is the infinite group $\mathfrak{C}_\infty \times \mathfrak{C}_\infty$ defined by 1.31 or $ab = ba$ or

$$aba^{-1}b^{-1} = E;$$

its two generators may be identified with the two circles that can be drawn in an obvious manner through an arbitrary point on the surface. For a surface of genus p, or sphere with p handles (Kerékjártó 1923, pp. 151—157; Threlfall 1932a, p. 33; Ball 1974, p. 233), it is the more general group

$$a_1b_1a_1^{-1}b_1^{-1}a_2b_2a_2^{-1}b_2^{-1} \ldots a_pb_pa_p^{-1}b_p^{-1} = E, \qquad (3.51)$$

which has a pair of generators for each handle. (This group is infinite for $p \geqq 1$, since it has a factor group $\mathfrak{C}_\infty \times \mathfrak{C}_\infty$ given by setting $a_k = b_k = E$ for every $k > 1$.) A more symmetrical definition for the same group is

$$A_1 A_2 \ldots A_{2p} A_1^{-1} A_2^{-1} \ldots A_{2p}^{-1} = E \tag{3.52}$$

(LEFSCHETZ 1949, p. 85).

The above surfaces are orientable. The simplest non-orientable surface is the projective plane, which may be regarded as a sphere with antipodes identified, or as a hemisphere with opposite peripheral points identified, or as a sphere with a cross-cap (HILBERT and COHN-VOSSEN 1932, p. 279). Its fundamental group is the $\mathfrak{C}_2$ defined by

$$A^2 = E,$$

whose single generator may be identified with a straight line of the projective plane. For a *Klein bottle* (or non-orientable torus, or sphere with two cross-caps) the group is

$$A_1^2 A_2^2 = E, \tag{3.53}$$

which is infinite since it has a factor group $\mathfrak{D}_\infty$ defined by $A_1^2 = A_2^2 = E$. For a sphere with q cross-caps (DEHN 1912, p. 131; LEFSCHETZ 1949, pp. 73—78) it is

$$A_1^2 A_2^2 \ldots A_q^2 = E, \tag{3.54}$$

with a generator for each cross-cap. (This is infinite for all $q > 1$, since an infinite factor group is obtained by setting $A_k = E$ for $k > 2$.)

After cutting an orientable or non-orientable surface along a set of m circuits,

$$m = 2p \text{ or } q,$$

which all pass through one point and generate the fundamental group, we can unfold it into a flat region bounded by a $2m$-gon whose sides,

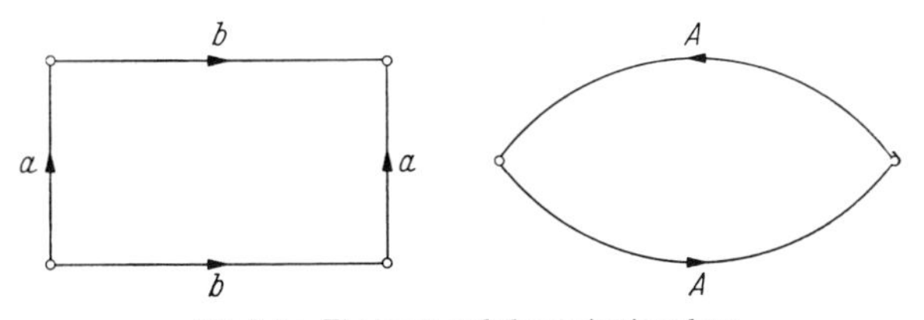

Fig. 3.5a. The torus and the projective plane

in their natural order, represent the abstract definition of the fundamental group in terms of these generators (LEFSCHETZ 1949, p. 165). For instance, the torus is unfolded into a rectangle, and the projective plane into a digon, as in Fig. 3.5a. (See also Fig. 3.5c for the cases $p = 2$ and $q = 3$.) By taking replicas of this polygon, one for each element of the fundamental group, and joining them along corresponding sides in the proper sense, we obtain the *universal covering surface*, which is simply-connected (THRELFALL 1932a, p. 8; HILBERT and COHN-VOSSEN 1932, p. 289). For instance, an infinity of rectangles fill the ordinary plane, as in Fig. 3.5b, and two digons (regarded as hemispheres) fill the sphere.

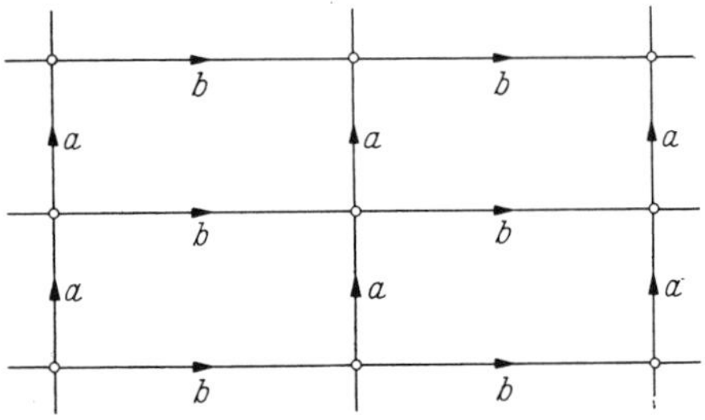

Fig. 3.5b. The fundamental group of the torus

The above remarks show that, on the simply-connected covering surface, *the sides of the polygons, marked as generators of the fundamental group, form a Cayley diagram for that group*.

If we introduce a metric so as to be able to use congruent polygons with straight sides, we must face the fact that, since each vertex belongs to $2m$ $2m$-gons, the sum of the angles of the polygon is just 2π. This means that the appropriate plane is elliptic if $q = 1$, Euclidean if $p = 1$ or $q = 2$, and hyperbolic if $p > 1$ or $q > 2$.

We have seen that the only *finite* fundamental groups are those of the sphere and the projective plane. This is the topological counterpart of the geometrical fact that the sphere and the elliptic plane have finite area, whereas the Euclidean and hyperbolic planes have infinite area. (For a remarkable drawing of the case $p = 2$, see HILBERT and COHN-VOSSEN 1932, p. 228, Abb. 249.)

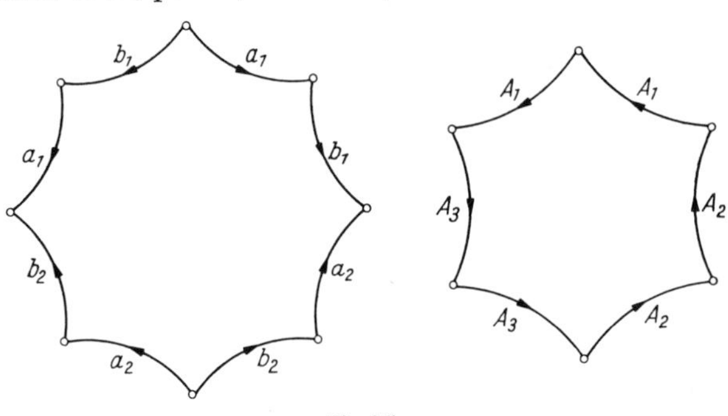

Fig. 3.5c

Along with the map of $2m$-gons on the universal covering surface, it is convenient to consider the *dual* map (on the same surface) which has a vertex in each face of the original map, an edge crossing each edge, and a face surrounding each vertex. The marking of edges can be carried over, and by directing them suitably we again have a Cayley diagram for the fundamental group. Fig. 3.5d shows a face of the original map and the corresponding vertex of the dual (in the cases $p = 2$ and $q = 3$). The multiply-connected surface is given by identifying pairs of sides of the face as indicated in Fig. 3.5c. The radial lines

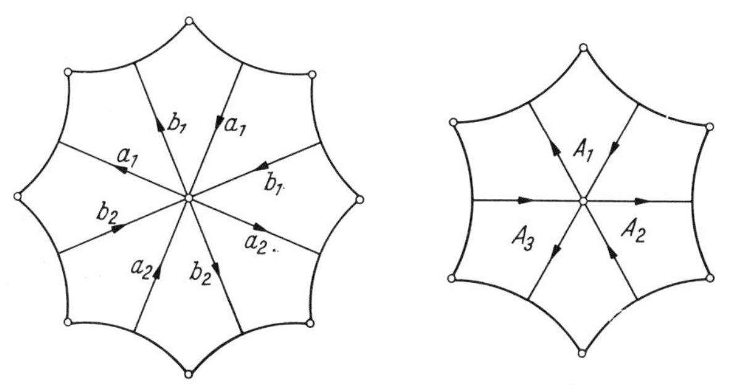

Fig. 3.5d

of Fig. 3.5d combine in pairs to form paths joining corresponding points on such pairs of sides. The identification makes these paths into circuits which can be used as generators of the fundamental group. Finally, as there is no need for these generators to pass through a single point, we can separate them as in Fig. 3.5e. (For an instance with $p > 2$, see HILBERT and COHN-VOSSEN 1932, p. 285, Abb. 322.)

Instead of dualizing the map on the universal covering surface, we can just as well dualize the corresponding map on the original surface. This is a map having one face, one vertex, and m edges (along which

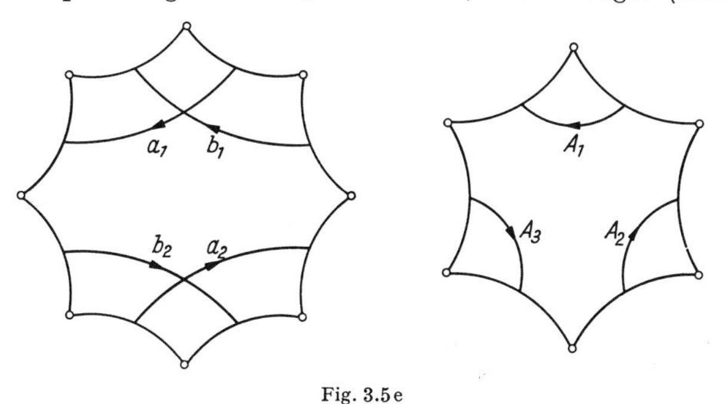

Fig. 3.5e

the surface was cut). Incidentally, the existence of this one-faced map shows that the characteristic of the surface is

$$N_0 - N_1 + N_2 = 2 - m \qquad (m = 2p \text{ or } q). \tag{3.55}$$

We conclude that, when the surface has been unfolded into a polygon whose sides are marked so as to be identified in pairs, the fundamental group has one set of generators represented by these pairs of sides and another set of generators represented by paths (inside the polygon) joining corresponding points on these pairs of sides (NIELSEN 1927, p. 195).

A more detailed account of fundamental groups will be given in Chapter 5.

3.6 Non-planar diagrams. We are now ready to investigate a Cayley diagram embedded into a surface of maximal characteristic, in the general case when this characteristic may be less than 2. Let the surface be unfolded by cutting it along a set of circuits that generate the fundamental group without passing through any vertices of the diagram. In other words, let the diagram be drawn inside a $2m$-gon with the understanding that an edge going out through any side is to be regarded as coming in again through the other side similarly marked, as in Fig. 3.6c.

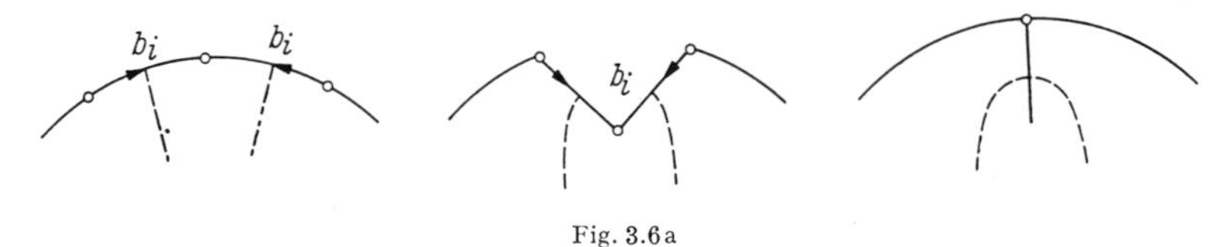

Fig. 3.6a

Every side of the polygon is crossed in this manner. For otherwise the same diagram could be drawn in a polygon having fewer sides, i.e., on a surface having a smaller p or q (and thus a greater characteristic). In the non-orientable case this is immediately clear, since the removal of a pair of sides of the $2q$-gon leaves a $2(q-1)$-gon of the same kind. In the orientable case (see 3.51), the removal of the pair of sides marked a_i (or b_i) leaves a pair marked b_i (or a_i) which are now adjacent. Since they are oppositely directed, they can be "cancelled out" by a simple distortion, as in Fig. 3.6a.

Since any path joining corresponding points on corresponding sides represents a generator of the fundamental group $\mathfrak{F}$ of the surface, it follows that each generator of $\mathfrak{F}$ is expressible as a word in the group $\mathfrak{G}$ whose Cayley diagram is under consideration. Also, every circuit in the diagram, being a circuit on the surface, is expressible as a word in $\mathfrak{F}$. Among the words C_k of 3.41, those represented by circuits round the faces (and trivial circuits along any undirected edges) enable us to equate the words represented by two homotopic circuits; in particular, they enable us to equate two expressions for a generator of $\mathfrak{F}$. Hence *the remaining words C_k may be identified with the m generators of $\mathfrak{F}$, and are given by paths joining corresponding sides of the $2m$-gon.*

This is the group-theoretical counterpart of the topological theorem that, when a graph is embedded into a surface of characteristic $2-m$, to form a map having N_2 faces, a set of μ fundamental circuits for the graph may be taken to consist of N_2-1 faces and the m generators of the fundamental group of the surface, in agreement with 3.11 and 3.55 (SEIFERT and THRELFALL 1947, § 46).

As before, the number of relations 3.41 may be reduced if the embedding is symmetrical, since we can then restrict attention to the faces surrounding a single vertex.

For example, consider the graph of Fig. 3.3b, which is a Cayley diagram for the quaternion group (§ 1.7, p. 8). It may be symmetrically embedded into the torus, as in Fig. 3.6b, where the second version

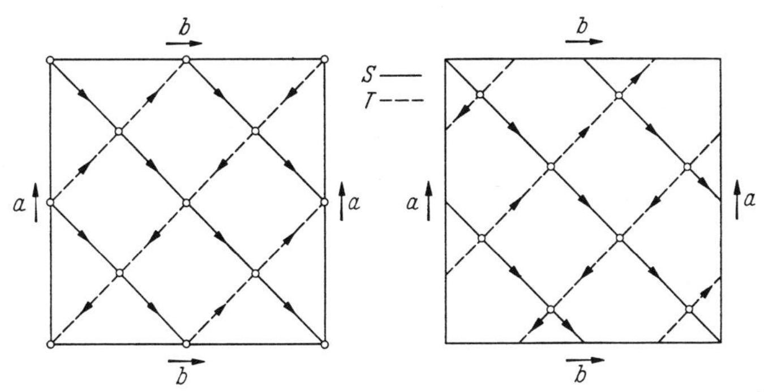

Fig. 3.6b. The quaternion group

avoids cutting through vertices. The map consists of 8 square faces, 4 at each vertex (COXETER 1950, pp. 414, 415; Figs. 4, 6). The faces that surround a vertex yield the four relations

$$TSTS^{-1} = STS^{-1}T = S^{-1}TST$$
$$= TS^{-1}TS = E,$$

which are all equivalent to the single relation

$$S^2 = (ST)^2;$$

but this relation does not suffice to define the group $\mathfrak{Q}$. Both generators of the fundamental group (one joining the vertical sides marked a, and one joining the horizontal sides marked b) may be expressed as S^2T^{-2}. Hence the required extra relation is

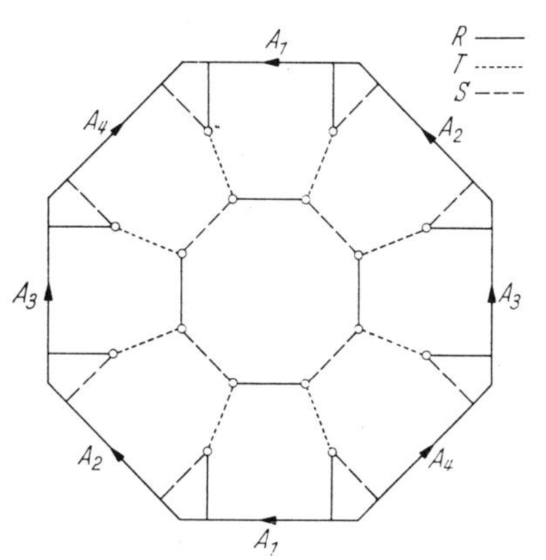

Fig. 3.6c. A group of order 16

$$S^2T^{-2} = E,$$

yielding the complete definition

$$S^2 = T^2 = (ST)^2.$$

Fig. 3.6c shows the graph of Fig. 3.3c (a Cayley diagram for a certain group of order 16) symmetrically embedded into the surface of genus 2. Here, for the sake of symmetry, we have unfolded the surface according to the scheme 3.52 instead of 3.51, so that the sides of the octagon to be identified are the pairs of *opposite* sides. The map has six octagonal faces, three at each vertex. As the generators are involutory, the edges are not directed. The faces and edges at a vertex yield the relations

$$R^2 = S^2 = T^2 = (ST)^4 = (TR)^4 = (RS)^4 = E.$$

The fundamental group now has four generators, one joining each pair of opposite sides. Two of them may be expressed as $TSRSTR$, and the

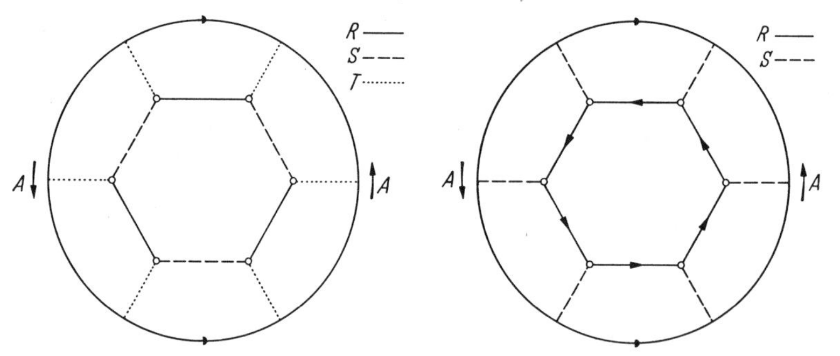

Fig. 3.6d. The non-cyclic group of order 6 Fig. 3.6e. The cyclic group of order 6

other two as $TRSRTS$. Equating these to E, we obtain altogether a redundant set of relations which is actually equivalent to

$$R^2 = S^2 = T^2 = E, \; RST = STR = TRS; \qquad (3.61)$$

for, this simple set of relations implies

$$(ST)^4 = STS \cdot TST \cdot ST = RTR \cdot RSR \cdot ST = (STR)^{-1} RST = E$$

and similarly $(TR)^4 = (RS)^4 = E$.

For the group $\mathfrak{D}_3$, the Cayley diagram Fig. 3.3d can be symmetrically embedded into the projective plane as in Fig. 3.6d. Here the map consists of one hexagonal and three quadrangular faces. Each vertex is incident with the hexagonal face and two quadrangular faces; these faces and the undirected edges at a vertex yield the relations

$$R^2 = S^2 = T^2 = (RS)^3 = RTST = STRT = E.$$

The single generator of the fundamental group may be taken to join any two opposite points on the peripheral circle (or "digon"). The consequent extra relation

$$SRST = E$$

yields $T = SRS$, so that R and S generate the group, and a set of

defining relations is
$$R^2 = S^2 = (RS)^3 = E.$$

Fig. 3.6e shows the same graph as a Cayley diagram of the cyclic group $\mathfrak{C}_6$, likewise embedded into the projective plane. The three faces at a vertex, together with the undirected edge, yield the relations
$$R^6 = S^2 = RSR^{-1}S = SR^{-1}SR = E.$$
In this case, the extra relation
$$R^3S = E$$
enables us to eliminate S, so that we are left with the usual definition 1.15.

3.7 Schreier's coset diagrams. The Cayley diagram for a given group $\mathfrak{G}$ is a graph whose vertices represent the elements of $\mathfrak{G}$, which are the

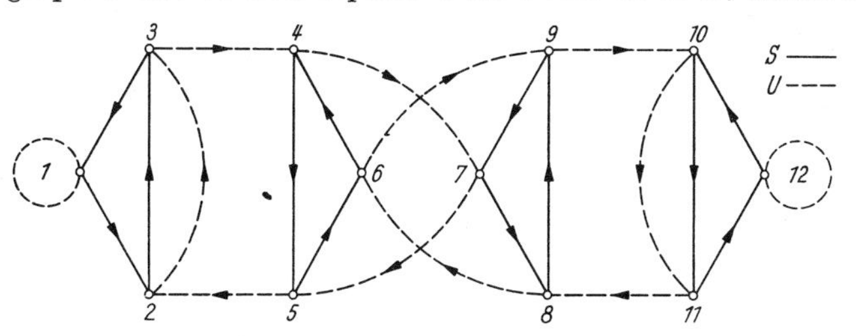

Fig. 3.7a. The icosahedral group

cosets of the trivial subgroup $\{E\}$. SCHREIER (1927, p. 180) generalized this notion by considering a graph whose vertices represent the cosets of any given subgroup $\mathfrak{H}$. This is essentially a graphical representation for

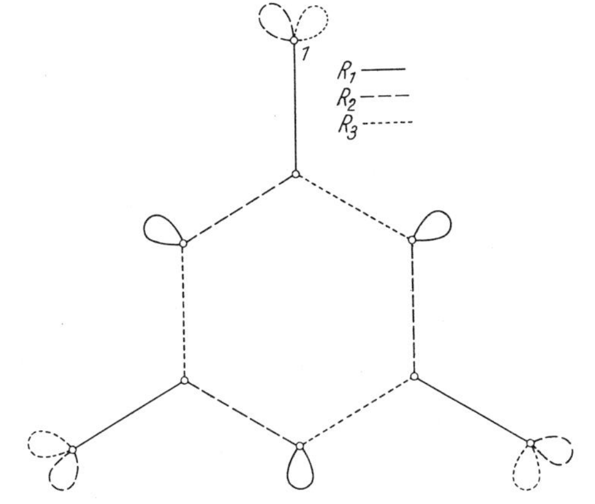

Fig. 3.7b. The group $[III]^3$ of order 54

the systematic enumeration of cosets (Chapter 2). The nodes representing cosets v and w are joined by an S_i-edge, directed from v to w, whenever
$$vS_i = w. \qquad (3.71)$$

It may well happen that $vS_i = v$, in which case the v-node is joined to itself by an S_i-loop. Comparing 3.71 with 3.31, we see that the Cayley diagram arises as the special case where $\mathfrak{H}$ is $\{E\}$, except that, for the sake of agreement with Chapter 2, we now read the word for a given path from left to right. (Thus we are considering the *right* coset diagram.)

The coset diagram for 2.21, using the cyclic subgroup $\{U\}$, is shown in Fig. 3.7a. Examples where the generators are involutory are

$$R_1^2 = R_2^2 = R_3^2 = (R_2 R_3)^3 = (R_3 R_1)^3 = (R_1 R_2)^3 = (R_1 R_2 R_3 R_2)^3 = E \quad (3.72)$$

(Fig. 3.7b; see COXETER **1957**b, p. 248), and

$$R_1^2 = R_2^2 = R_3^2 = (R_2 R_3)^3 = (R_3 R_1)^3 = (R_1 R_2)^3 = E \quad (3.73)$$

(Fig. 3.7c), using in both cases the dihedral subgroup $\{R_2, R_3\}$, of order 6.

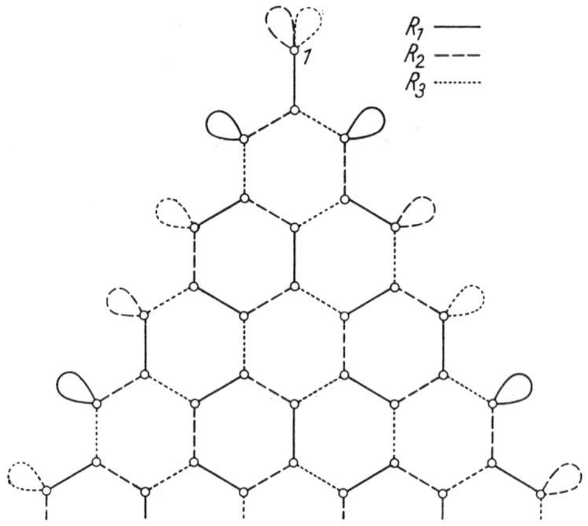

Fig. 3.7c. The infinite group $\triangle$

In the former case there are 9 nodes, representing 9 cosets, and the order is 54. In the latter, the pattern continues indefinitely, showing that the group is infinite. (It is the group $\triangle$ of COXETER 1963a, p. 79.)

Chapter 4

Abstract Crystallography

When we say that a square is more symmetrical than a triangle, we mean that it has a greater number of *symmetry operations*, i.e., of congruent transformations that leave it invariant. The symmetry operations (including the identity) of any figure clearly form a group: the *symmetry group* of the figure. A completely irregular figure has a symmetry group of order 1. The group of order 2 arises when the figure has bilateral symmetry, or when it is transformed into itself by a half-turn (i.e., rotation through two right angles).

We apologize for using $\{\}$ in two different senses: the polyhedron or tessellation $\{p, q\}$ consisting of p-gons, q at each vertex, and the group $\{R, S\}$ generated by R and S.

4.1 The cyclic and dihedral groups. A rotation through $2\pi/q$ evidently generates the cyclic group $\mathfrak{C}_q$; for instance, $\mathfrak{C}_4$ is the group of the swastika, and $\mathfrak{C}_5$ of the periwinkle flower, *vinca herbacea* (WEYL 1952, p. 66). Combining $\mathfrak{C}_q$ with the reflection in a line through the centre of rotation, we obtain the dihedral group $\mathfrak{D}_q$, of order $2q$, which we thus identify with the complete symmetry group of the regular q-gon, $\{q\}$; for instance, $\mathfrak{D}_5$ is the group of a geranium flower. WEYL attributes to LEONARDO DA VINCI the discovery that $\mathfrak{C}_q$ and $\mathfrak{D}_q$ are the only finite two-dimensional symmetry groups. Similarly, $\mathfrak{C}_1$ and $\mathfrak{D}_1$ are the only finite one-dimensional groups, while $\mathfrak{C}_\infty$ and $\mathfrak{D}_\infty$ are the only infinite discrete one-dimensional groups.

The infinite discrete two-dimensional groups arise in practice as the various ways of repeating a flat design, as on wallpaper or a tiled floor. All 17 of them were discovered empirically by the Moors in their decoration of the Alhambra in Granada; many of them also by the ancient Egyptians and the Chinese (SPEISER 1924, pp. 86—90; DYE 1937; MÜLLER 1944; WEYL 1952, pp. 104—115; WELLS 1956, p. 83). We shall obtain abstract definitions for them in § 4.5.

Strictly, the number of infinite two-dimensional groups is not 17 but 24. However, to save space, we omit the seven groups containing translations in only one direction, i.e., the symmetry groups of patterns on a strip (SPEISER 1924, pp. 82—83; COXETER 1963b, p. 71).

4.2 The crystallographic and non-crystallographic point groups. The finite three-dimensional symmetry groups were first enumerated in 1830 by HESSEL, whose work remained unnoticed until EDMUND HESS arranged for its republication in OSTWALD's *Klassiker* in 1897. The essential ideas are as follows.

Every finite group of congruent transformations leaves invariant at least one point: the centroid (barycentre) of all the transforms of an arbitrary point. Every congruent transformation that leaves a point invariant is either the reflection in a plane through the point or the product of two or three such reflections (COXETER 1963a, p. 36). It preserves or reverses sense according as the number of reflections is even or odd. Hence every sense-preserving transformation (with a fixed point) is the product of just two reflections, i.e., it is a *rotation* (through twice the angle between the reflecting planes). The invariant point enables us to regard the group as operating on a sphere.

One special sense-reversing transformation is the central inversion Z, which reverses every vector and thus interchanges antipodes on the sphere. It is the product of reflections in three mutually orthogonal planes. Combining Z with any other sense-reversing transformation, we obtain the product of an even number of reflections, i.e., a rotation. Thus every sense-reversing transformation is a *rotatory inversion*: the product of a rotation and the central inversion.

Since every sense-preserving transformation (with a fixed point) is a rotation, the product of two rotations (about concurrent axes) is another rotation, and *every finite group of sense-preserving transformations is a group of rotations.* A simple argument shows that the only groups of this kind are the rotational symmetry groups of:

1) the q-gonal pyramid,
2) the q-gonal dipyramid or prism,
3) the regular tetrahedron $\{3, 3\}$,
4) the cube $\{4, 3\}$ or octahedron $\{3, 4\}$,
5) the dodecahedron $\{5, 3\}$ or icosahedron $\{3, 5\}$

(Ford 1929, p. 133; Zassenhaus **1958**, pp. **16—19**; Weyl 1952, pp. 77—79, 149—154; Coxeter 1963b, § 15.4). The complete symmetry group of each figure includes reflections, but for the moment we consider the subgroup consisting of rotations alone.

We immediately recognize 1) and 2) as the cyclic and dihedral groups, $\mathfrak{C}_q$ and $\mathfrak{D}_q$. The complete symmetry group of the tetrahedron is the symmetric group on its four vertices; hence the subgroup 3), of order 12, is the alternating group $\mathfrak{A}_4$. The rotations of the cube permute its four diagonals (joining pairs of opposite vertices); hence 4), of order 24, is the symmetric group $\mathfrak{S}_4$.

The twenty vertices of the dodecahedron can be distributed among five inscribed tetrahedra, as in Fig. 4.2 (cf. Hess 1876, p. 45). Each rotation permutes these five tetrahedra evenly; hence 5), of order 60, is the alternating group $\mathfrak{A}_5$.

Because of these geometric representations, the groups $\mathfrak{A}_4$, $\mathfrak{S}_4$ and $\mathfrak{A}_5$ are usually called the *tetrahedral, octahedral* and *icosahedral* groups (Klein 1876; 1884, pp. 14—19).

It was observed by Pólya and Meyer (1949; see also Weyl 1952, pp. 120, 155—156) that every other finite symmetry group is either the direct product of a rotation group with the $\mathfrak{C}_2$ generated by Z, or else the following kind of combination of two rotation groups $\mathfrak{G}$ and $\mathfrak{H}$, where $\mathfrak{H}$ is a subgroup of index 2 in $\mathfrak{G}$. We take the elements of $\mathfrak{H}$ as they stand, and multiply each of the remaining elements of $\mathfrak{G}$ by Z. The resulting group, which has the same order as $\mathfrak{G}$, is $\mathfrak{H}[\mathfrak{G}$ in Pólya's notation, $\mathfrak{G}\mathfrak{H}$ in Weyl's. In Table 2 (on page 135) we give also the classical notations of SCHOENFLIES (1891, p. 146) and of HERMANN and MAUGUIN (HENRY and LONSDALE 1952, pp. 25, 44).

The 32 groups whose rotations all have periods 2, 3, 4 or 6 are called *crystallographic* point groups, or crystal *classes* (see, e.g., BURCKHARDT

Fig. 4.2. Five tetrahedra whose vertices belong to a dodecahedron

1947, p. 72), because they occur as factor groups of infinite discrete symmetry groups or *space groups*. They were classified as abstract groups by BELOWA, BELOW and SCHUBNIKOW (1948). The remaining (non-crystallographic) point groups were discussed by DOLIVO-DOBROVOLSKY (1925). NOWACKI (1933) showed which groups occur as subgroups in others.

4.3 Groups generated by reflections. The groups generated by reflections deserve special consideration for two reasons: there is a general theory covering them all, and they contain the remaining point groups as subgroups.

The group generated by reflections in any number of planes is equally well generated by reflections in all their transforms: a configuration of planes that is symmetrical by reflection in each one. If the group is finite, the planes all pass through the invariant point (or points) and determine a corresponding configuration of great circles on a sphere. These great circles decompose the sphere into a finite number of regions (namely hemispheres, lunes, or spherical triangles) whose angles are submultiples of π (COXETER 1963a, pp. 76–77). All the regions are congruent (with a possible reversal of sense), since each reflects into its neighbours.

The simplest instance is the group [1], of order 2, generated by the reflection in a single plane which cuts the sphere into two hemispheres. When a group is generated by two reflections, we can take the angle between the reflecting planes to be π/q $(q \geqq 2)$. The planes and their transforms meet the sphere in a pencil of q meridians which decompose it into $2q$ lunes. The Cayley diagram is a $2q$-gon with one vertex in each lune. Thus we have the group

$$[q] \cong \mathfrak{D}_q,$$

of order $2q$, with the abstract definition

$$R_1^2 = R_2^2 = (R_1 R_2)^q = E. \tag{4.31}$$

Other such groups are generated by reflections R_1, R_2, R_3 in the three sides of a spherical triangle with angles π/p_{23}, π/p_{31}, π/p_{12}, say. The reflections clearly satisfy

$$R_1^2 = R_2^2 = R_3^2 = E$$

and three relations of the form $(R_i R_j)^{p_{ij}} = E$. The following remarks show that every relation satisfied by the R's is an algebraic consequence of these.

Calling the initial triangle "region E", we observe that any element S of the group transforms it into a congruent "region S". In particular the generators transform region E into the neighbouring regions R_i; the element S transforms E with its neighbours R_i into S with its neighbours $R_i S$. Thus we pass through the ith side of region S to enter region $R_i S$. Any expression for S as a word

$$\ldots R_k R_j R_i$$

corresponds to a path from a position inside region E to a position inside region S, passing through the ith side of E, then through the jth side of R_i, through the kth side of $R_j R_i$, and so on (reading from right to left[1]). Two different paths from E to S can be combined to form a closed path from E to $S^{-1}S = E$, corresponding to a word that is equal to E. Since the sphere is simply-connected, such a closed path can be decomposed into elementary circuits of two kinds: one going from S to a neighbouring region $R_i S$ and back to $R_i^2 S = S$, and one going round a vertex where $2p_{ij}$ regions meet, from S to $(R_i R_j)^{p_{ij}} S = S$. A gradual shrinkage of the path corresponds to an algebraic reduction

[1]) In the words of a letter from A. Speiser (December 1954):
Wenn man ein Produkt von Substitutionen geometrisch deutet, muß man sie „raumfest" deuten, falls man von links nach rechts liest. Liest man aber von rechts nach links, so muß man sie „körperfest" deuten.

of the word to E by means of the relations $R_i^2 = E$ and $(R_i R_j)^{p_{ij}} = E$. It follows that these relations suffice for an abstract definition, and that the triangle is a *fundamental region*: there is one such triangle for each element of the group; and any point on the sphere, being inside or on the boundary of such a region S, is derivable from a corresponding point of region E by the transformation S.

The triangles covering the sphere may be regarded as the faces of a map (§ 3.2). Naming points within them, instead of the faces themselves, we obtain the dual map, whose edges cross the sides of the triangles. The above remarks serve to identify this dual map with the Cayley diagram (§ 3.3), which accordingly has, at each vertex, a $2p_{23}$-gon, a $2p_{31}$-gon, and a $2p_{12}$-gon, representing the relations

$$(R_2 R_3)^{p_{23}} = (R_3 R_1)^{p_{31}} = (R_1 R_2)^{p_{12}} = E.$$

Since the fundamental region is a spherical triangle, its angle-sum must exceed π; hence

$$\frac{1}{p_{23}} + \frac{1}{p_{31}} + \frac{1}{p_{12}} > 1.$$

Since $\frac{1}{3} + \frac{1}{3} + \frac{1}{3} = 1$, the smallest of the p's must be 2, and the others, say p and q, satisfy

$$\frac{1}{p} + \frac{1}{q} > \frac{1}{2}, \quad \text{or} \quad (p-2)(q-2) < 4.$$

We thus obtain the cases $[2, q]$, $[3, 3]$, $[3, 4]$, $[3, 5]$ of the group

$$[p, q] \text{ or } [q, p],$$

defined by

$$R_1^2 = R_2^2 = R_3^2 = (R_1 R_2)^p = (R_2 R_3)^q = (R_3 R_1)^2 = E, \qquad (4.32)$$

whose fundamental region is a triangle with angles π/p, π/q, $\pi/2$. This group is the complete symmetry group of either of the two reciprocal regular polyhedra $\{p, q\}$, $\{q, p\}$ (COXETER 1963a, p. 83). The Cayley diagram consists of the vertices and edges (not directed) of the semiregular polyhedron

$$t\begin{Bmatrix} p \\ q \end{Bmatrix}$$

(COXETER 1940a, p. 394) which has, at each vertex, a $2p$-gon, a $2q$-gon, and a square. (The case $[3, 5]$ is illustrated in Fig. 4.3, which resembles a drawing by R. FRICKE, see PASCAL 1927, p. 945. The classical symbols for all these groups are given in Table 2.)

Since the area of the fundamental region is measured by its angular excess, the order of $[p, q]$ is equal to the number of such triangles that will be needed to cover the whole sphere, namely

$$\frac{4}{\frac{1}{p}+\frac{1}{q}-\frac{1}{2}}=\frac{8pq}{4-(p-2)(q-2)}$$

(COXETER 1963a, p. 82).

Fig. 4.3. The group $[3, 5] \cong \mathfrak{C}_2 \times \mathfrak{A}_5$

4.4 Subgroups of the reflection groups. The three rotations

$$R=R_1R_2, S=R_2R_3, T=R_3R_1,$$

or any two of them, generate a "polyhedral" subgroup $[p, q]^+$ or $[q, p]^+$ of order $4pq\{4 - (p-2)(q-2)\}^{-1}$, defined by

$$R^p = S^q = T^2 = RST = E \tag{4.41}$$

or

$$R^p = S^q = (RS)^2 = E \tag{4.42}$$

or

$$S^q = T^2 = (ST)^p = E \tag{4.43}$$

(see COXETER 1940a, where $[p, q]^+$ was called $[p, q]'$).

The concept of a fundamental region remains valid; but since the generators do not leave its sides invariant, the shape of the region is no longer uniquely determined. Since $[p, q]^+$ is of index 2 in $[p, q]$, the area of its fundamental region is double that of the triangle considered above. It is natural to choose such a shape that the network of congruent regions and the Cayley diagram form dual maps (BURNSIDE 1911, pp. 406, 423). In the case of 4.43, we combine two adjacent triangles so as to form a larger triangle with two angles π/p and one $2\pi/q$, and the Cayley diagram is $\mathrm{t}\{p, q\}$ (which has, at each vertex, two $2p$-gons and a q-gon). In the case of 4.42 with $q > 2$, we again combine two of the right-angled triangles, but now we choose a pair that have a common hypotenuse, so as to obtain a kite-shaped quadrangle, and the Cayley diagram is $\mathrm{r}\begin{Bmatrix} p \\ q \end{Bmatrix}$ (which has, at each vertex, two non-adjacent squares separated by a p-gon and a q-gon). In the case of 4.41, we combine one triangle with portions of its three neighbours (say a white triangle with portions of the three adjacent black triangles) to form a pentagon (or, if $p = 2$, a quadrangle), and the Cayley diagram is $\mathrm{s}\begin{Bmatrix} p \\ q \end{Bmatrix}$. (See

BURNSIDE 1911, frontispiece, for the case $[3, 4]^+$. Unfortunately the direction of his arrows disagrees with the text on pp. 424, 427. The operation $S_1S_2S_3$ should take us along an S_3-edge, then an S_2-edge, and then an S_1-edge.)

Nearly all these Cayley diagrams were given by MASCHKE (1896, pp. 156—194, Figs. 2—10, 16—18; cf. R. P. BAKER 1931, pp. 645—646; COXETER, LONGUET-HIGGINS and MILLER 1954, pp. 403, 439, Figs. 15—25, 27, 29—32).

When $p = 3$ and $q = 4$ or 5, the element $(R_1R_2R_3)^{2q-5}$ of $[p, q]$ is the central inversion Z (COXETER 1963a, p. 91). It follows that $[p, q]$ is then the direct product of the group $\{Z\}$ of order 2 and the rotation subgroup $[p, q]^+$.

When q is even, $[p, q]$ has another subgroup of index 2, say

$$[p^+, q] \text{ or } [q, p^+],$$

generated by the rotation $R = R_1R_2$ and the reflection R_3. We easily derive from 4.32 the abstract definition

$$R^p = R_3^2 = (R^{-1}R_3RR_3)^{q/2} = E \qquad (4.44)$$

(COXETER 1940a, p. 387). The fundamental region is most naturally taken to be a triangle with two angles π/q and one $2\pi/p$ (or, if $p = 2$, a lune), and the Cayley diagram is $t\{q, p\}$ (or, if $p = 2$, $\{2q\}$). This resembles the diagram for $[p, q]^+$ in the form

$$R^p = T^2 = (RT)^q = E,$$

differing only in the way the various R-edges are directed. When $q = 2$, we have $[p^+, 2]$ or $[2, p^+]$, the direct product of the $\mathfrak{C}_p$ generated by R and the $\mathfrak{C}_2$ generated by R_3.

The most interesting group of this kind is $[3^+, 4]$, the direct product of the $\mathfrak{C}_2$ generated by the central inversion $(RR_3)^3$ and the $[3, 3]^+$ generated by R and R_3RR_3. It is the symmetry group of a dodecahedron with an inscribed cube, or of an octahedron with an inscribed icosahedron (COXETER 1940a, p. 396), or of the crystallographic *pyritohedron*.

When $p = 2$, the relations 4.44 reduce to

$$R^2 = R_3^2 = (RR_3)^q = E \qquad (q \text{ even}).$$

In this dihedral group $[2^+, q]$, the rotatory-inversion RR_3 generates a cyclic subgroup of order q. Pursuing the convention whereby each superscript $^+$ halves the order, we denote this subgroup by

$$[2^+, q^+] \qquad (q \text{ even})$$

(see Table 2 on page 135). It is, of course, a subgroup of index 4 in $[2, q]$, and is generated by $R_1R_2R_3$. We note for comparison that, when q is odd, this element of $[2, q]$ generates $[2, q^+]$ (of index 2).

4.5 The seventeen two-dimensional space groups. The first mathematical treatment of the plane crystallographic groups was by FEDOROV (1891). They were rediscovered by FRICKE and KLEIN (1897, pp. 227—233) and again by PÓLYA (1924) and NIGGLI (1924; see also BURCKHARDT 1947, pp. 111—124; HEESCH and KIENZLE 1963, pp. 23—29). NOWACKI (1954, p. 133) showed them to be abstractly distinct. In this section we shall obtain an abstract definition for each group in terms of two, three or four

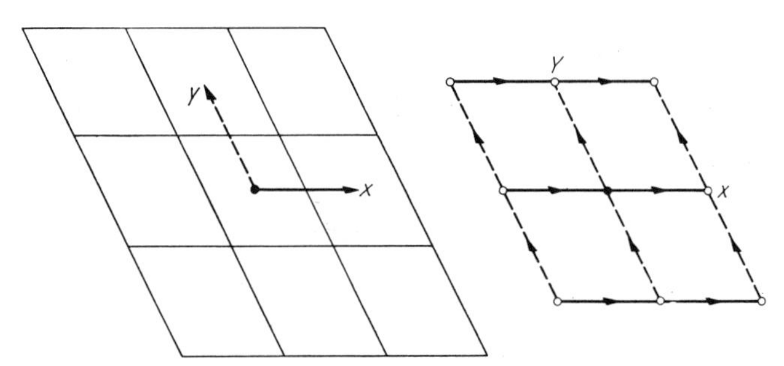

Fig. 4.5a. **p 1** generated by two translations

generators, observing that the network of fundamental regions (indicated in the left half of each figure) and the Cayley diagram (in the right half) form dual maps. The point E of the Cayley diagram is marked as a black dot inside the fundamental region on the left. We denote half-turns and other rotations by the customary symbols, translations by arrows, and glide-reflections by "half arrows".

In referring to the individual groups, it seems simplest to use the "short" symbols of HERMANN and MAUGUIN (HENRY and LONSDALE 1952, pp. 46—72). The connection with other symbols is shown in Table 3.

Any two independent translations (or vectors) X, Y in the plane generate a lattice whose translation group **p 1** is defined abstractly by the relation 1.31:

$$XY = YX \qquad (4.501)$$

(Fig. 4.5a). Since this group is the direct product of the free groups $\{X\}$ and $\{Y\}$, it may be described as $\mathfrak{C}_\infty \times \mathfrak{C}_\infty$. Its general element $X^x Y^y$ (where x and y are integers) is the translation from the origin to the point with affine coordinates (x, y).

The fundamental region may be chosen in infinitely many ways, but if we insist that it be convex, it must have a centre of symmetry (FEDOROV 1885, pp. 271—272) and in fact must be either a parallelogram (as above) or a centrally symmetrical hexagon. The latter possibility arises when we use the three generators X, Y, $Z = X^{-1}Y^{-1}$ to generate the same group in the form

$$XYZ = ZYX = E \tag{4.5011}$$

(Fig. 4.5b).

By changing the magnitudes and directions of the translations X, Y, we obtain infinitely many geometrical varieties of the single abstract group. This multiplicity could be avoided by regarding the plane as being affine instead of Euclidean, so that all parallelograms are alike (COXETER 1955, p. 11).

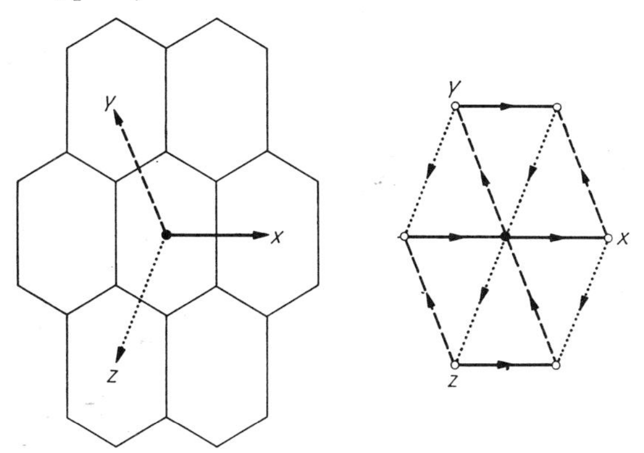

Fig. 4.5b. **p 1** generated by three translations

Any half-turn T reverses the directions of the translations X, Y, which generate **p 1**. The group **p 2**, obtained by adjoining this T to **p 1**, is defined by the relations

$$XY = YX,\ T^2 = E,\ TXT = X^{-1},\ TYT = Y^{-1}.$$

The three half-turns $T_1 = TY$, $T_2 = XT$, $T_3 = T$ generate this larger group, and in terms of them its defining relations are

$$T_1^2 = T_2^2 = T_3^2 = (T_1 T_2 T_3)^2 = E \tag{4.502}$$

(Fig. 4.5c), or, in terms of T_1, T_2, T_3, $T_4 = T_1 T_2 T_3 = T_1 X$,

$$T_1^2 = T_2^2 = T_3^2 = T_4^2 = T_1 T_2 T_3 T_4 = E \tag{4.5021}$$

(Fig. 4.5d; see also DYCK 1882, p. 37).

It is noteworthy that *any* triangle will serve as a fundamental region for **p 2** in the form 4.502, the three generators being half-turns about the mid-points of the three sides. This is obvious when we use the affine plane, in which all triangles are alike. Similarly, any simple quadrangle (not necessarily convex) will serve as a fundamental region for the same group **p 2** in the form 4.5021, the four generators being half-turns about the mid-points of the four sides (STEINHAUS 1950, p. 60, Fig. 56; FEJES TÓTH 1953, p. 66, Abb. 61).

The remaining fifteen groups require the concept of right angle, and are therefore not affine but strictly Euclidean. However, some of them still involve a free parameter, yielding an infinity of geometrical varieties for one abstract group.

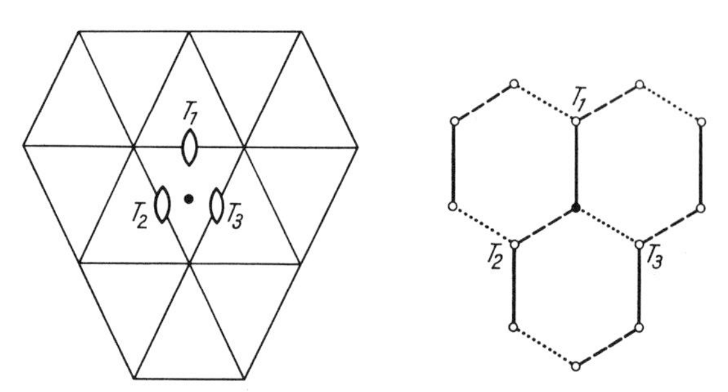

Fig. 4.5c. **p 2** generated by three half-turns

When the translations X, Y which generate **p 1** are in perpendicular directions, the reflection R in a line parallel to Y transforms X into X^{-1} and leaves Y invariant. Hence the group **p m**, obtained by adjoining R to **p 1**, is defined by the relations

$$XY = YX, \; R^2 = E, \; RXR = X^{-1}, \; RYR = Y.$$

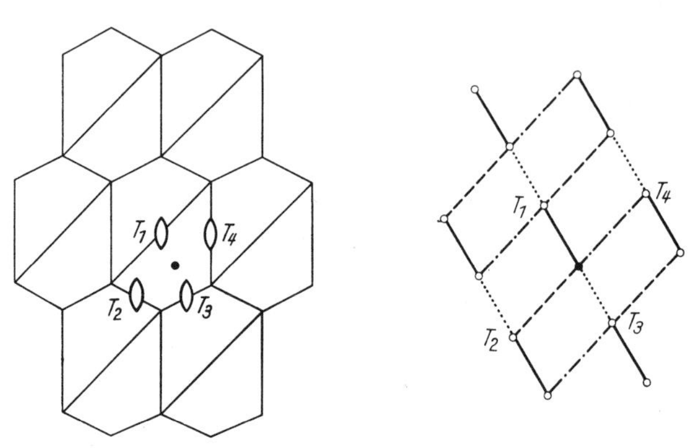

Fig. 4.5d. **p 2** generated by four half-turns

In terms of Y, R, and the reflection $R' = RX$, these relations become

$$R^2 = R'^2 = E, \; RY = YR, \; R'Y = YR' \qquad (4.503)$$

(Fig. 4.5e), so that **p m** is the direct product $\{R, R'\} \times \{Y\}$, and may be described as $\mathfrak{D}_\infty \times \mathfrak{C}_\infty$.

Again, when the translations X and Y of **p 1** are in perpendicular directions, there is a glide-reflection P, whose square is Y, which transforms X into X^{-1}. If we adjoin P to **p 1** we obtain the group **p g** defined by the relations

$$XY = YX,\quad P^2 = Y,\quad P^{-1}XP = X^{-1}.$$

In terms of the two parallel glide-reflections P, $Q = PX$, this group is defined by the single relation

$$P^2 = Q^2 \tag{4.504}$$

(Fig. 4.5f), and may therefore be described as $\langle 2, 2, \infty \rangle$ in the notation of 1.64.

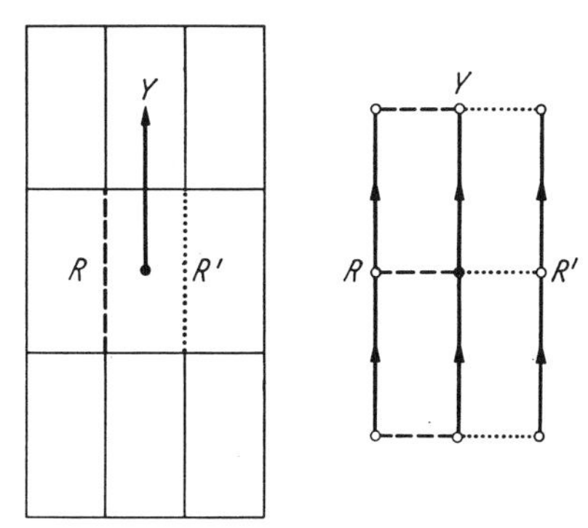

Fig. 4.5e. **p m** generated by two reflections and a translation

This group **p g** shares with **p l** (in the form $XY = YX$) the property that its generators leave no point invariant. It is therefore the fundamental group (§ 3.5, p. 24) for a closed surface of characteristic 0, derived from the fundamental region by identifying certain pairs of sides. In the case of **p l**, defined by $XY = YX$ or $XYX^{-1}Y^{-1} = E$, the fundamental region is a parallelogram; we identify the two opposite sides related by X, and the two opposite sides related by Y, to obtain the *torus*. In the case of **p g**, where the fundamental region is a rhombus,

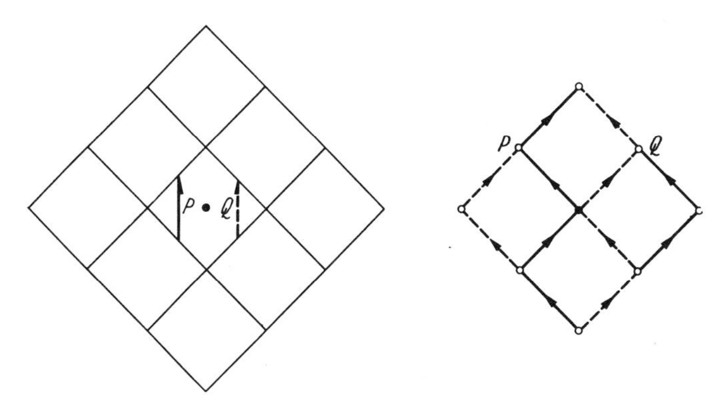

Fig. 4.5f. **p g** generated by two parallel glide-reflections

we identify the two adjacent sides related by P, and the two adjacent sides related by Q, to obtain the *Klein bottle*. To transform the relation $P^2 = Q^2$ into the canonical form 3.53, we merely have to write $A_1 = P$, $A_2 = Q^{-1}$.

If we adjoin to **p g** the reflection R which interchanges P and Q, we obtain the group **c m** defined by the relations

$$P^2 = Q^2,\ R^2 = E,\ RPR = Q.$$

Eliminating Q, we have

$$R^2 = E,\ RP^2 = P^2R \tag{4.505}$$

(Fig. 4.5g). In terms of R and the translation $S = PR$, the same group **c m** is defined by

$$R^2 = E,\ (RS)^2 = (SR)^2.$$

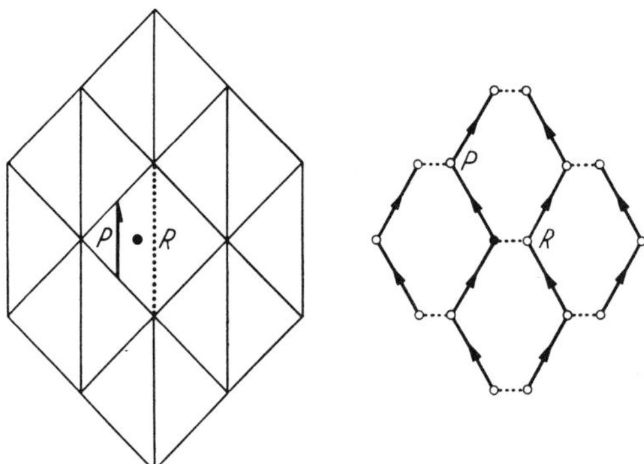

Fig. 4.5g. **c m** generated by a reflection and a glide-reflection

There is a reflection R_2 which leaves the reflections R and R' of **p m** invariant, while transforming the translation Y into its inverse. The group **p m m**, obtained by adjoining this R_2 to **p m**, is defined by the relations

$$R^2 = R'^2 = R_2^2 = E,\ RY = YR,\ R'Y = YR',$$
$$R_2RR_2 = R,\ R_2R'R_2 = R',\ R_2YR_2 = Y^{-1}$$

or, in terms of the four reflections $R_1 = R$, R_2, $R_3 = R'$, $R_4 = R_2Y$,

$$R_1^2 = R_2^2 = R_3^2 = R_4^2 = (R_1R_2)^2 = (R_2R_3)^2 = (R_3R_4)^2 = (R_4R_1)^2 = E. \tag{4.506}$$

This, being the direct product of $\{R_1, R_3\}$ and $\{R_2, R_4\}$, may be described as $\mathfrak{D}_\infty \times \mathfrak{D}_\infty$.

In Fig. 4.5h, the fundamental region for **p m m** has been drawn as a square. But an isomorphic group is obtained by using any rectangle. However, since the generators leave invariant the sides of the rectangle (one by one), we cannot cut off a part of the fundamental region and compensate by adding a congruent piece elsewhere, as one can when other kinds of generators occur. This possibility of varying the shape of the fundamental

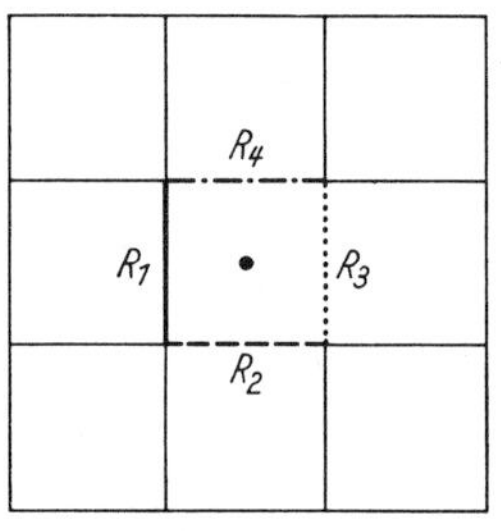

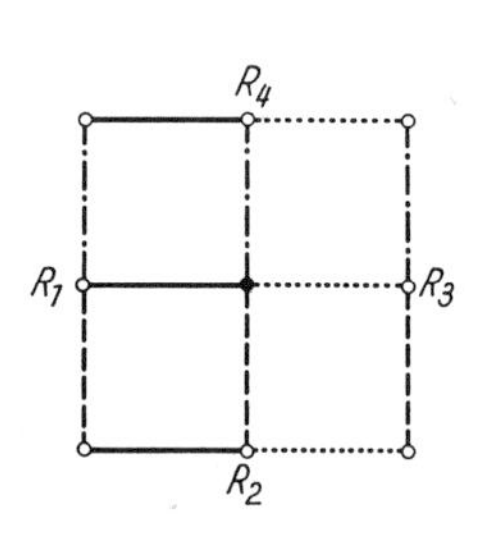

Fig. 4.5h. **p m m** generated by four reflections

region (for a group not entirely generated by reflections) has been neatly employed by the Dutch artist M. C. ESCHER, who draws the fundamental region of **p g** as a horseman, that of **c m** as a beetle, and so on (ESCHER 1961; COXETER 1963b, § 4.3).

There is a reflection R that reverses the directions of the glide-reflections P and Q which generate **p g**. Extending **p g** by this reflection, we obtain the group **p m g**, defined by the relations

$$P^2 = Q^2,\ R^2 = E,\ RPR = P^{-1},\ RQR = Q^{-1}$$

or, in terms of R and the half-turns $T_1 = PR$, $T_2 = QR$,

$$R^2 = T_1^2 = T_2^2 = E,\ T_1RT_1 = T_2RT_2 \tag{4.507}$$

(Fig. 4.5i).

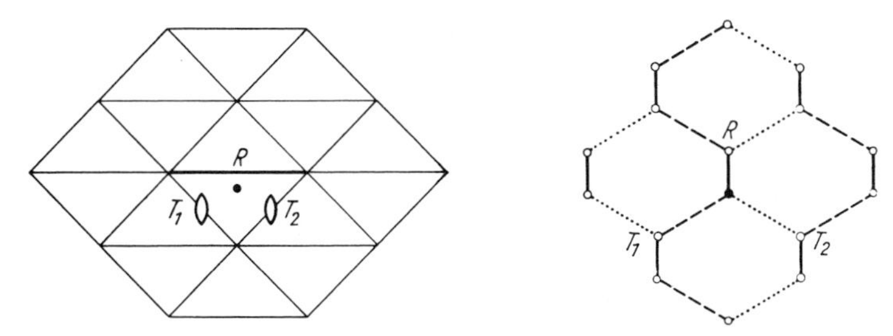

Fig. 4.5i. **p m g** generated by a reflection and two half-turns

Adjoining to the group **p g** the half-turn T which transforms P into Q^{-1} and Q into P^{-1}, we obtain the group **p g g** defined by the relations

$$P^2 = Q^2,\ T^2 = E,\ TPT = Q^{-1}$$

or, in terms of the perpendicular glide-reflections $O = PT$ and P,

$$(PO)^2 = (P^{-1}O)^2 = E \tag{4.508}$$

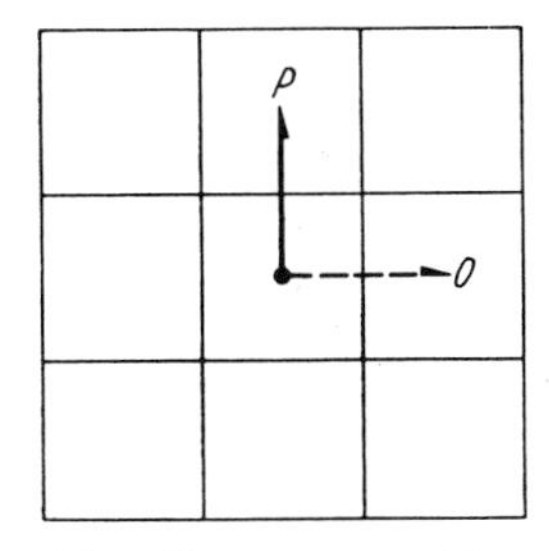

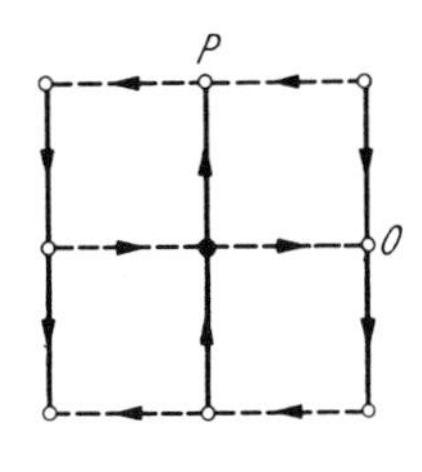

Fig. 4.5j. **p g g** generated by two perpendicular glide-reflections

(Fig. 4.5j). Thus **p g g** is $(\infty, \infty \mid 2, 2)$ in the notation of COXETER (1939, pp. 74, 81).

There is a half-turn T that interchanges the pairs of opposite reflections R_1, R_3 and R_2, R_4 which generate **p m m**. The group **c m m**, obtained by adjoining this T to **p m m**, is defined by the relations

$$R_1^2 = R_2^2 = R_3^2 = R_4^2 = (R_1 R_2)^2 = (R_2 R_3)^2 = (R_3 R_4)^2 = (R_4 R_1)^2 = E,$$
$$T^2 = E,\ T R_1 T = R_3,\ T R_2 T = R_4$$

or, in terms of the two reflections R_1, R_3, and the half-turn T,

$$R_1^2 = R_2^2 = T^2 = (R_1 R_2)^2 = (R_1 T R_2 T)^2 = E. \qquad (4.509)$$

It is of special interest as being the symmetry group of the "anomalous" uniform tessellation (COXETER 1940a, p. 405; STEINHAUS 1950, p. 65, Fig. 65) in which strips of squares alternate with strips of triangles.

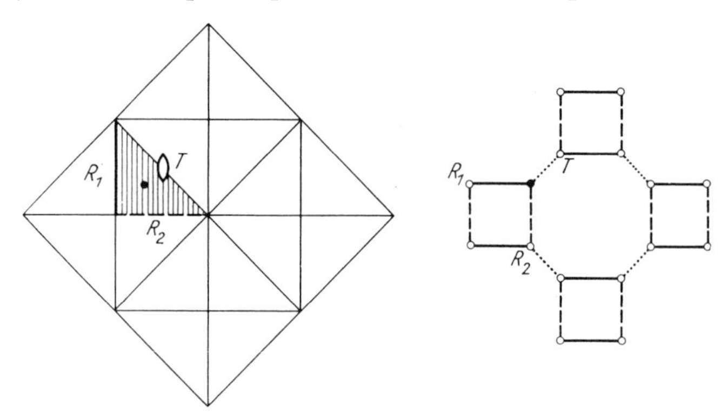

Fig. 4.5k. **c m m** generated by two reflections and a half-turn

If the fundamental region of **p 2** is taken to be a square, there is a tetragonal rotation S that cyclically permutes the half-turns T_1, T_2, T_3, T_4. The group **p 4**, obtained by adjoining this rotation to **p 2**, is defined by the relations

$$T_1^2 = T_2^2 = T_3^2 = T_4^2 = T_1 T_2 T_3 T_4 = E,$$
$$S^4 = E,\ S^{-i} T_4 S^i = T_i \qquad (i = 1, 2, 3)$$

or, in terms of $T = T_4$ and S,

$$S^4 = T^2 = (S T)^4 = E \qquad (4.510)$$

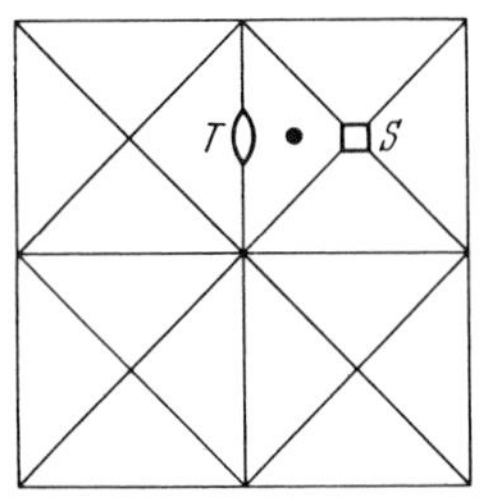

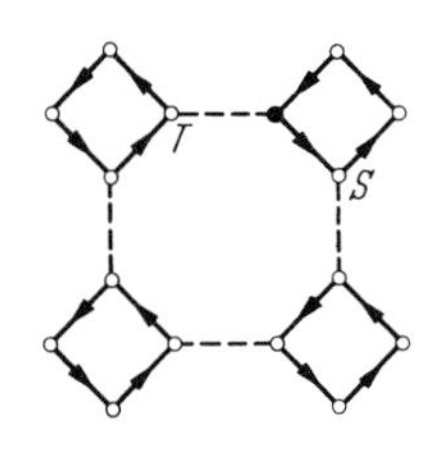

Fig. 4.51. **p 4** generated by a tetragonal rotation and a half-turn

(Fig. 4.51). Thus **p 4** is $[4, 4]^+$ in the notation of 4.43. (See also BURNSIDE 1911, p. 416. For the same group in the form $R^4 = S^4 = (RS)^2 = E$, see COXETER 1948, p. 24, Fig. 11.)

The fundamental region of **p m m** may be any rectangle. If it is taken to be a square, there is a reflection R which interchanges R_1 and R_4, and interchanges R_2 and R_3. The group **p 4 m**, obtained by adjoining this reflection to **p m m**, is defined by the relations

$$R_1^2 = R_2^2 = R_3^2 = R_4^2 = (R_1 R_2)^2 = (R_2 R_3)^2 = (R_3 R_4)^2 = (R_4 R_1)^2 = E,$$
$$R^2 = E, \quad R R_1 R = R_4, \quad R R_2 R = R_3$$

or, in terms of R, R_1, R_2,

$$R^2 = R_1^2 = R_2^2 = (R R_1)^4 = (R_1 R_2)^2 = (R_2 R)^4 = E \qquad (4.511)$$

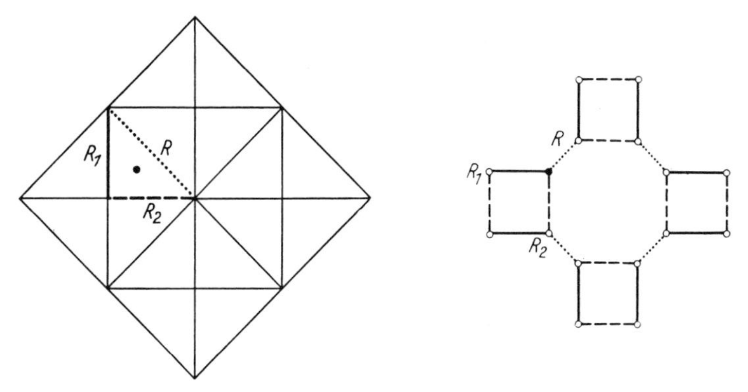

Fig. 4.5m. **p 4 m** generated by reflections in the sides of a right-angled isosceles triangle

(Fig. 4.5m). Thus **p 4 m** is $[4, 4]$ in the notation of 4.32; i.e., it is the complete symmetry group of the regular tessellation of squares, $\{4, 4\}$ (COXETER 1963a, p. 59).

Again, when the fundamental region of **p m m** is taken to be a square, there is a tetragonal rotation S that cyclically permutes the reflections R_1, R_2, R_3, R_4. The group **p 4 g**, obtained by adjoining this rotation to

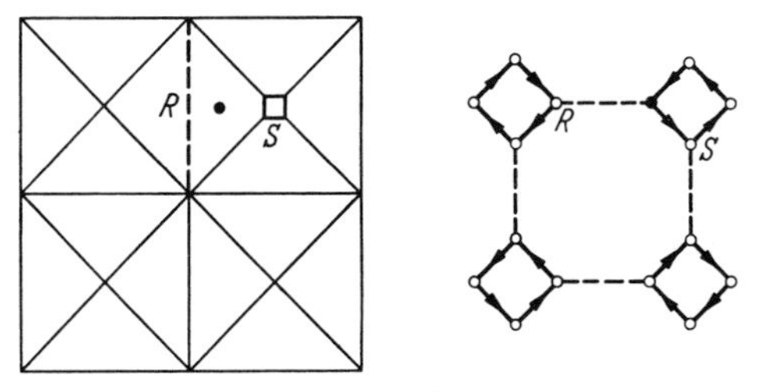

Fig. 4.5n. **p 4 g** generated by a reflection and a tetragonal rotation

p m m, is defined by the relations

$$R_1^2 = R_2^2 = R_3^2 = R_4^2 = (R_1R_2)^2 = (R_2R_3)^2 = (R_3R_4)^2 = (R_4R_1)^2 = E,$$
$$S^4 = E, \quad S^{-i}R_4S^i = R_i \qquad (i = 1, 2, 3)$$

or, in terms of $R = R_4$ and S,

$$R^2 = S^4 = (RS^{-1}RS)^2 = E \tag{4.512}$$

(Fig. 4.5n). Thus **p 4 g** is $[4^+, 4]$ in the notation of 4.44 (see also COXETER 1963a, p. 66). It is the symmetry group of the uniform tessellation $s\{4, 4\} = s\begin{Bmatrix}4\\4\end{Bmatrix}$ (COXETER 1940a, p. 394; STEINHAUS 1950, p. 64, Fig. 63).

When the translations (or vectors) X and Y, which generate **p 1**, are equal in length, while the angle between them is $2\pi/3$, the extra generator $Z = (XY)^{-1}$ has this same length and makes the same angle with both. Hence there is a trigonal rotation S that cyclically permutes the translations X, Y, Z. The group **p 3**, obtained by adjoining this rotation to **p 1**, is defined by the relations

$$XYZ = ZYX = E,\ S_1^3 = E,\ S_1^{-1}XS_1 = Y,\ S_1^{-1}YS_1 = Z,\ S_1^{-1}ZS_1 = X.$$

In terms of the trigonal rotations S_1, $S_2 = S_1X$, $S_3 = X^{-1}S_1$, these relations become

$$S_1^3 = S_2^3 = S_3^3 = S_1S_2S_3 = E \tag{4.513}$$

(Fig. 4.5o), or, in terms of S_1 and S_2,

$$S_1^3 = S_2^3 = (S_1S_2)^3 = E \tag{4.5131}$$

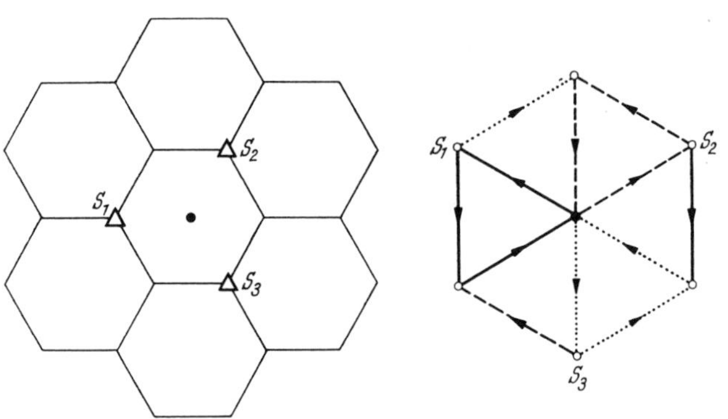

Fig. 4.5o. **p 3** generated by three trigonal rotations

(Fig. 4.5p). Thus **p 3** is one of the groups discussed by BURNSIDE (1911, p. 143).

Adjoining to **p 3** the reflection R which transforms S_1 into S_2^{-1} and S_2 into S_1^{-1}, we obtain the group **p 3 1 m** defined by the relations

$$S_1^3 = S_2^3 = (S_1S_2)^3 = E, \quad R^2 = E, \quad RS_1R = S_2^{-1}$$

or, in terms of the reflection R and the rotation $S = S_1$,

$$R^2 = S^3 = (RS^{-1}RS)^3 = E \qquad (4.514)$$

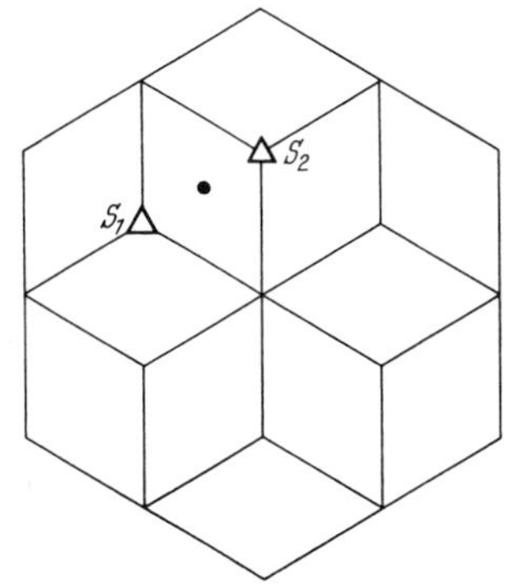

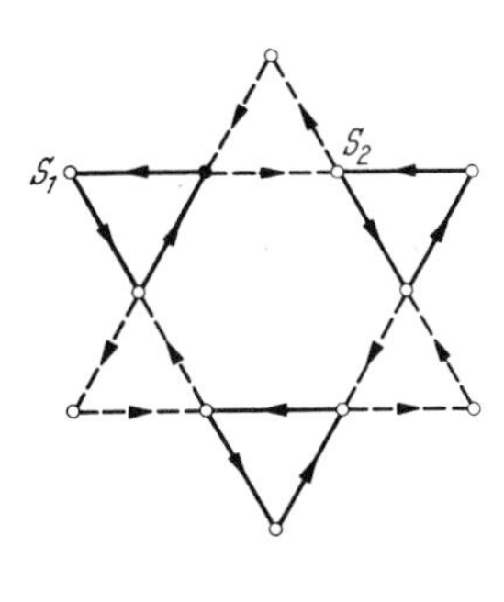

Fig. 4.5p. **p 3** generated by two trigonal rotations

(Fig. 4.5q). Thus **p 3 1 m** is $[3^+, 6]$ in the notation of 4.44 (see COXETER 1936a, p. 67).

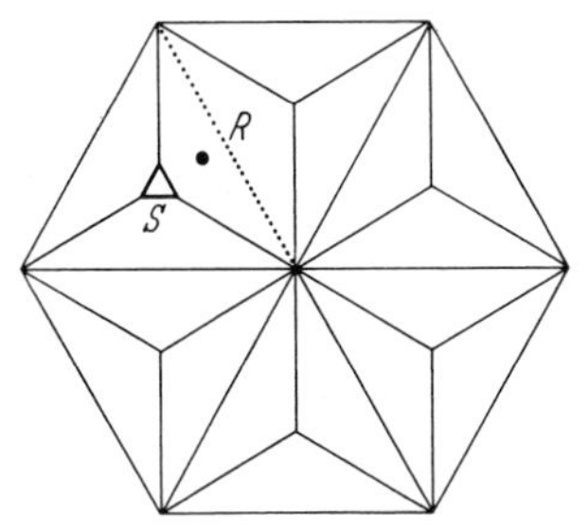

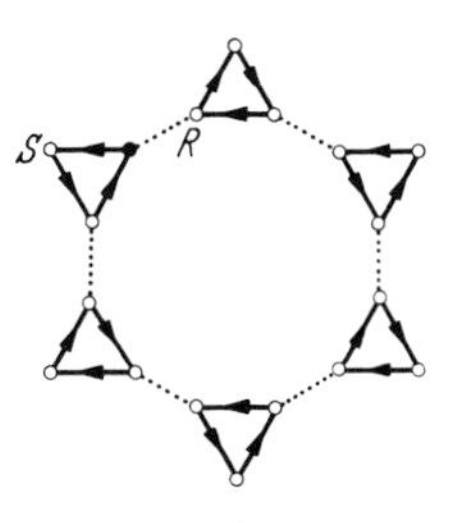

Fig. 4.5q. **p 3 1 m** generated by a reflection and a trigonal rotation

There is a reflection R that transforms the trigonal rotations S_1, S_2 of **p 3** into their inverses. The group **p 3 m 1**, obtained by adjoining this reflection to **p 3**, is defined by the relations

$$S_1^3 = S_2^3 = (S_1S_2)^3 = E,\ R^2 = E,\ RS_1R = S_1^{-1},\ RS_2R = S_2^{-1}$$

or, in terms of the reflections

$$R_1 = RS_2,\ R_2 = S_1R,\ R_3 = R,$$

$$R_1^2 = R_2^2 = R_3^2 = (R_1R_2)^3 = (R_2R_3)^3 = (R_3R_1)^3 = E \qquad (4.515)$$

(Fig. 4.5r; cf. 3.73).

There is a half-turn T that interchanges the trigonal rotations S_1 and S_2 of **p 3**. The group **p 6**, obtained by adjoining this half-turn to **p 3**, is defined by the relations

$$S_1^3 = S_2^3 = (S_1S_2)^3 = E,\ T^2 = E, TS_1T = S_2$$

or, in terms of the rotation $S = S_1$ and the half-turn T,

$$S^3 = T^2 = (ST)^6 = E \qquad (4.516)$$

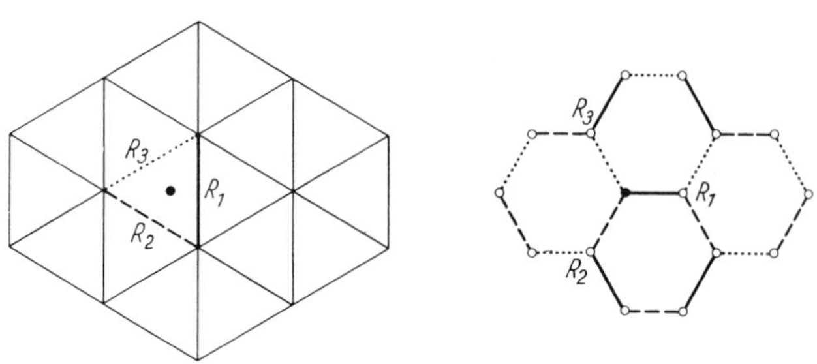

Fig. 4.5r. **p 3 m 1** generated by reflections in the sides of an equilateral triangle

(Fig. 4.5s). Thus **p 6** is $[3, 6]^+$ in the notation of 4.43 (see also BURNSIDE 1911, p. 417). It is the complete symmetry group of the uniform tessellation $s\left\{\begin{matrix}3\\6\end{matrix}\right\}$, which has four triangles and a hexagon at each vertex (COXETER 1940a, p. 392; STEINHAUS 1950, p. 65, Fig. 64); in fact, this is its Cayley diagram when **p 6** is expressed in the form

$$R^6 = S^3 = T^2 = RST = E.$$

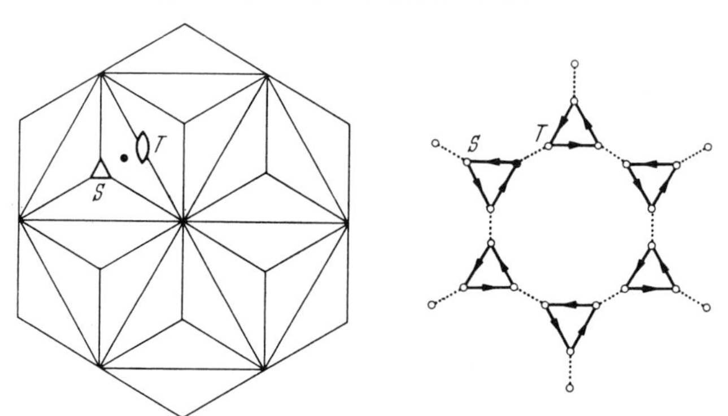

Fig. 4.5s. **p 6** generated by a trigonal rotation and a half-turn

There is a reflection R that interchanges the reflections R_1, R_3 of **p 3 m 1**, and leaves the reflection R_2 invariant. The group **p 6 m**, obtained by adjoining this reflection to **p 3 m 1**, is defined by the relations

$$R_1^2 = R_2^2 = R_3^2 = (R_1R_2)^3 = (R_2R_3)^3 = (R_3R_1)^3 = E,$$

$$R^2 = E, \quad RR_1R = R_3, \quad RR_2R = R_2$$

or, in terms of the three reflections R, R_1, R_2,

$$R^2 = R_1^2 = R_2^2 = (R_1R_2)^3 = (R_2R)^2 = (RR_1)^6 = E \qquad (4.517)$$

(Fig. 4.5t). Thus **p 6 m** is $[6, 3]'$ or $[3, 6]$, in the notation of 4.32; i.e., it is the complete symmetry group of either of the dual regular tessellations $\{6, 3\}$, $\{3, 6\}$, formed by hexagons and triangles, respectively.

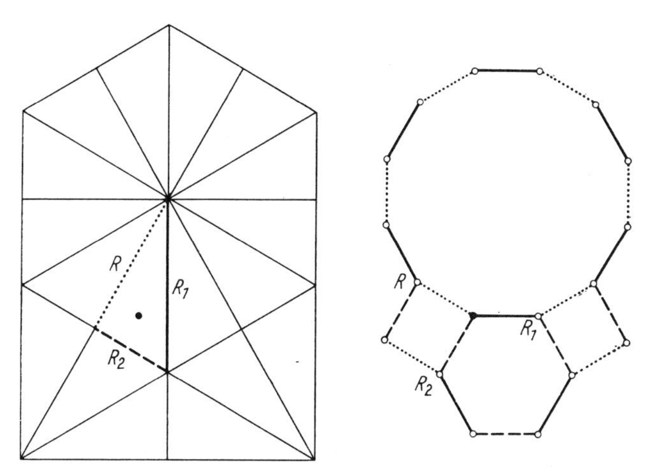

Fig. 4.5t. **p 6 m** generated by reflections in the sides of a bisected equilateral triangle

4.6 Subgroup relationships among the seventeen groups. NIGGLI (1924, p. 297) indicated many ways in which these groups occur as subgroups of one another. However, since his incomplete table was reproduced by HENRY and LONSDALE (1952, p. 537), we have undertaken an exhaustive investigation, resulting in Table 4. There is one row and one column for each group, and each entry is the index of the column-group as a subgroup in the row-group, in ordinary type when it is a normal subgroup, *italic* type when it is not. In the main diagonal we give the (smallest) index of each group as a proper subgroup of itself.

The details of the work are too lengthy to be given here. (They may be seen in MOSER's Ph. D. dissertation in the Library of the University of Toronto.) However, it seems worth while to show how the two most complicated groups, **p 4 m** and **p 6 m**, together contain all the others as subgroups.

In **p 4 m** $\cong [4, 4]$, defined by 4.511,

R_1 and $S = RR_2$	generate **p 4 g** $\cong [4^+, 4]$,
S and $T_1 = R_1R_2$	generate **p 4** $\cong [4, 4^+]$,
$R, R' = R_2S$ and T_1	generate **c m m**,
$P = R_1S$ and $O = SR_1$	generate **p g g**,

R and P generate **c m**,
P and $Q = RPR$ generate **p g**,
$R_4 = RR_1R$, T_1, and $T_2 = S^2$ generate **p m g**,
T_1, T_2, and $T_3 = RT_1R$ generate **p 2**,
R_1, R_2, $R_3 = SR$ and R_4 generate **p m m** $\cong [\infty] \times [\infty]$,
R_1, R_3, and $Y = R_2R_4$ generate **p m** $\cong [\infty] \times [\infty]^+$,
$X = R_1R_3$ and Y generate **p 1** $\cong [\infty]^+ \times [\infty]^+$.

In **p 6 m** $\cong [3, 6]$, defined by 4.517,

R and $S = R_1R_2$ generate **p 3 1 m** $\cong [3^+, 6]$,
S and $T = R_2R$ generate **p 6** $\cong [3, 6]^+$,
R_1, R_2 and $R_3 = RR_1R$ generate **p 3 m 1** $\cong \triangle$,
$S_1 = R_1R_2$ and $S_2 = R_2R_3$ generate **p 3** $\cong \triangle^+$.

In the last column of Table 3, the symbols for the generators have been slightly altered (for simplicity, and to stress certain analogies).

Chapter 5

Hyperbolic Tessellations and Fundamental Groups

After describing the regular tessellations $\{p, q\}$ (which are simply-connected topological polyhedra), we shall derive the fundamental groups (§ 3.5, p. 24) for closed surfaces of arbitrary characteristic. From the present standpoint, non-orientable surfaces are easier than orientable surfaces; accordingly, we consider them in this unusual order.

5.1 Regular tessellations. In the Euclidean plane, the angle of a regular p-gon, $\{p\}$, is $(1 - 2/p)\,\pi$; hence q equal $\{p\}$'s (of any size) will fit together round a common vertex if this angle is equal to $2\pi/q$, that is, if

$$(p-2)\,(q-2) = 4.$$

We thus obtain the three regular tessellations $\{p, q\}$, namely

$$\{4, 4\}, \{3, 6\}, \{6, 3\}.$$

The angle of a *spherical* $\{p\}$ is *greater* than $(1 - 2/p)\,\pi$, and gradually increases from this value to π when the circum-radius increases from 0 to $\pi/2$. Hence, if

$$(p-2)\,(q-2) < 4,$$

we can adjust the size of the polygon so that the angle is exactly $2\pi/q$; then q such $\{p\}$'s will fit together round a common vertex, and we have the "spherical tessellations" $\{p, q\}$, namely

$$\{2, q\}, \{q, 2\}, \{3, 3\}, \{3, 4\}, \{4, 3\}, \{3, 5\}, \{5, 3\}. \tag{5.11}$$

The first of these, formed by q lunes joining two antipodal points, is the q-gonal *hosohedron*; the second, formed by two q-gons (each covering a hemisphere), is called the q-gonal *dihedron*. The rest are "blown up" variants of the five Platonic solids (§ 4.2).

In the hyperbolic plane, the angle of a $\{p\}$ is less than $(1 - 2/p)\,\pi$, and gradually decreases from this value to zero when the circum-radius increases from 0 to ∞. Hence, if

$$(p-2)\,(q-2) > 4,$$

we can adjust the size of the polygon so that the angle is exactly $2\pi/q$; then q such $\{p\}$'s will fit together round a common vertex, and we can continue to add further $\{p\}$'s indefinitely. In this manner we construct a *hyperbolic tessellation* $\{p, q\}$, which is an infinite collection of regular p-gons filling the whole hyperbolic plane (SCHLEGEL 1883, p. 360. Note that our $\{p, q\}$ is the $\{q, p\}$ of THRELFALL 1932a, p. 32).

The tessellation $\{p, q\}$ is symmetrical by reflection in certain lines, which may be of as many as three different types: in-radii of faces, circum-radii of faces, and edges. If p or q is odd, some lines will play two of these roles; if both are odd, every line of symmetry plays all three roles (COXETER 1963a, pp. 65, 86). Reflections in these lines satisfy 4.32 and generate $[p, q]$, the complete symmetry group of $\{p, q\}$. The fundamental region is a triangle ONZ (see Fig. 5.2 for the case $p = q = 4$), where O is a vertex, N is the mid-point of an edge MO, and Z is the centre of a face KMO, so that the angles are π/p at Z, $\pi/2$ at N, and π/q at O (cf. COXETER 1963a, pp. 67, 90, where ONZ is called $P_0P_1P_2$).

The lines of the hyperbolic plane can be represented conformally by circles orthogonal to one circle in the Euclidean plane. A very artistic pattern is obtained by shading alternate replicas of the fundamental region. (See KLEIN and FRICKE 1890, p. 109 for the case of [3, 7]; MILLER 1911, p. 395 for [3, 8]; COXETER 1939, pp. 126, 127 for [4, 5] and [4, 6].)

5.2 The Petrie polygon. A *Petrie polygon* of $\{p, q\}$ is a "zigzag" $\ldots KMOQ \ldots$ (Fig. 5.2) in which every two consecutive sides, but no three, belong to a face (COXETER 1963a). A particular specimen of this "polygon" is determined by two adjacent sides of any face; therefore the various specimens are all alike, and we are justified in speaking of *the* Petrie polygon of the tessellation. The dual tessellation $\{q, p\}$ has a corresponding Petrie polygon whose sides cross those of $\{p, q\}$. In the case of a spherical tessellation, the Petrie polygon has h sides, where

$$\cos^2 \frac{\pi}{h} = \cos^2 \frac{\pi}{p} + \cos^2 \frac{\pi}{q} \tag{5.21}$$

(COXETER 1963a, p. 19).

We see from Fig. 5.2 that the sides of the Petrie polygon are permuted by the operation $R_1 R_2 R_3$, which takes KMO to MOQ. In the Euclidean and hyperbolic cases, this is a glide-reflection (COXETER 1947, p. 202) whose axis joins the mid-points $L, N, P, \ldots$ of the sides of the Petrie polygon; i.e., it is the product of the reflection in LN and the translation from L to N. In the hyperbolic case, the amount of this translation, being twice the distance from N to ZO, is given by

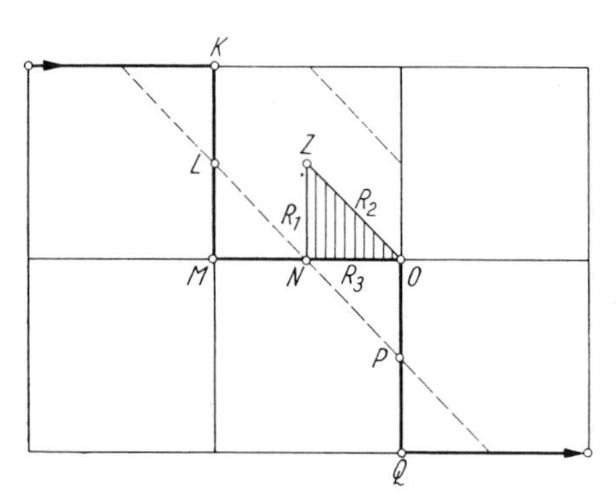

Fig. 5.2. A Petrie polygon of {4, 4}

$$\cosh^2 \frac{LN}{2} = \cos^2 \frac{\pi}{p} + \cos^2 \frac{\pi}{q}$$

(COXETER and WHITROW 1950, p. 422, with $r = 2$). By repeating the glide-reflection, we obtain a pure translation of amount $LP = 2LN$. In terms of the rotations $R = R_1 R_2$ and $S = R_2 R_3$, this is

$$(R_1 R_2 R_3)^2 = R_1 R_2 R_3 R_2 R_2 R_1 R_2 R_3 = R S^{-1} R^{-1} S.$$

Thus the various commutators of R and S have the effect of shifting the Petrie polygon, in various positions, two steps along itself.

5.3 Dyck's groups. Since the Euclidean, spherical and hyperbolic planes are simply-connected, the topological argument in § 4.3, p. 35, shows that the relations 4.32 provide an abstract definition for $[p, q]$. It follows that the relations 4.41 (or 4.42, or 4.43) provide an abstract definition for the "rotation" subgroup $[p, q]^+$. Since the Euclidean and hyperbolic planes are infinite, we have thus proved that

The group

$$R^p = S^q = (RS)^2 = E$$

is infinite whenever $(p-2)(q-2) \geq 4$.

This is surely one of the most remarkable contributions of geometry to algebra. For, the algebraic proof (MILLER 1902) is excessively complicated, requiring separate consideration of many different cases.

More generally, any p-gon with angles $\pi/q_1, \pi/q_2, \ldots, \pi/q_p$ can be repeated, by successive reflections in sides, to cover the sphere, or the Euclidean or hyperbolic plane; in other words, any such polygon will serve as fundamental region for a group generated by reflections. This group has the abstract definition

$$R_1^2 = R_2^2 = \cdots = R_p^2 = (R_1 R_2)^{q_1} = (R_2 R_3)^{q_2} = \cdots = (R_p R_1)^{q_p} = E,$$

and its "rotation subgroup" is given by

$$S_1^{q_1} = S_2^{q_2} = \cdots = S_p^{q_p} = S_1 S_2 \ldots S_p = E$$

(DYCK 1882, p. 28). Hence these groups are infinite whenever

$$\frac{1}{q_1} + \frac{1}{q_2} + \cdots + \frac{1}{q_p} \leqq p - 2$$

(THRELFALL 1932a, pp. 26—29). We may even include polygons having one or more vertices at infinity, by taking $1/q_i$ to be zero and omitting the corresponding relation. When $p > 3$, we can change the sides without changing the angles, but the corresponding groups are isomorphic; e.g., any rectangle may be used for **p m m** (4.506).

In particular, whenever q is even, a face of $\{p, q\}$ serves as fundamental region for the group

$$R_1^2 = R_2^2 = \cdots = R_p^2 = (R_1 R_2)^{q/2} = (R_2 R_3)^{q/2} = \cdots = (R_p R_1)^{q/2} = E. \quad (5.31)$$

Since the area of this p-gon is $2p$ times that of the fundamental region for $[p, q]$, 5.31 is a subgroup of index $2p$ in $[p, q]$. In terms of 4.32, the p generators are transforms of R_3 by powers of the rotation $R_1 R_2$, namely

$$(R_2 R_1)^{i-1} R_3 (R_1 R_2)^{i-1} \qquad (i = 1, 2, \ldots, p).$$

Transforming by R_1 and R_2 in turn, we see that 5.31 is a *normal* subgroup of $[p, q]$.

The simplest instance is the trirectangular (self-polar) spherical triangle, which is both a face of $\{3, 4\}$ and a fundamental region for the crystallographic point group **m m m** or $[2, 2]$, of order 8, defined by

$$R_1^2 = R_2^2 = R_3^2 = (R_2 R_3)^2 = (R_3 R_1)^2 = (R_1 R_2)^2 = E \qquad (5.32)$$

(see Table 2). This is a normal subgroup of index 6 in the extended octahedral group **m 3 m** or $[3, 4]$. Another instance is the Euclidean equilateral triangle, which is a face of $\{3, 6\}$ and a fundamental region for the two-dimensional space group **p 3 m 1** or $\triangle$, of index 6 in **p 6 m** or $[3, 6]$. (Table 4 gives 2 as the index of **p 3 m 1** in **p 6 m**; but each of these groups is of index 3 in itself, or rather, in a similar group.) Yet another instance is the square, which is a face of $\{4, 4\}$ and a fundamental region for **p m m** or $[\infty] \times [\infty]$, of index 8 in **p 4 m** or $[4, 4]$. Finally, an "extreme" instance is the trebly asymptotic triangle (having all three vertices at infinity), which is a face of $\{3, \infty\}$ and a fundamental region for the infinite group

$$R_1^2 = R_2^2 = R_3^2 = E.$$

This is a normal subgroup of index 6 in the extended modular group $[3, \infty]$ (KLEIN 1879a, pp. 120—121), which we shall discuss in § 7.2, p. 85.

In each case, the dual tessellation $\{q, p\}$, with edges suitably coloured (but not directed), is the Cayley diagram for the group 5.31.

5.4 The fundamental group for a non-orientable surface, obtained as a group generated by glide-reflections. Writing $2q$ for both p and q in 4.32, we see that the group $[2q, 2q]$, defined by

$$R_1^2 = R_2^2 = R_3^2 = (R_1 R_2)^{2q} = (R_2 R_3)^{2q} = (R_3 R_1)^2 = E,$$

has a normal subgroup of index $4q$, whose fundamental region is a face of $\{2q, 2q\}$. Each generator of the subgroup, being a reflection, leaves invariant every point on a line (the line containing a side of the $\{2q\}$). In this section we shall describe another subgroup with the same fundamental region (and therefore the same index, $4q$) but having the remarkably different property that its generators, and in fact all its elements except E, leave no point invariant.

One case is very simple: when $q = 1$, the point group **m m m** or $[2, 2]$, of order 8, defined by 5.32, has a subgroup **1** or $[2^+, 2^+]$, of order 2, generated by the central inversion $R_1 R_2 R_3$. Its fundamental region, being a hemisphere, is one of the two faces of the "digonal dihedron" $\{2, 2\}$.

When $q > 1$, we can transform one face of $\{2q, 2q\}$ into its neighbours by *glide-reflections*, forwards and backwards, along the segments joining the mid-points of q alternate pairs of adjacent sides. (See Fig. 5.2 for the case $q = 2$, Fig. 5.4a for $q = 3$.) Since each glide-reflection and its inverse yield two of the $2q$ neighbours, we are careful to select only q of the $2q$ pairs of adjacent sides. Therefore the selected glide-reflections will not generate a *normal* subgroup of $[2q, 2q]$ (except when $q = 1$).

As we saw in § 5.2, one such glide-reflection is $R_1 R_2 R_3$. To find them all, we transform by powers of the rotation $(R_1 R_2)^2$, obtaining

$$A_i = (R_2 R_1)^{2i} R_1 R_2 R_3 (R_1 R_2)^{2i} = (R_2 R_1)^{2i-1} R_3 (R_1 R_2)^{2i}. \quad (5.41)$$

Since

$$A_i^2 = (R_2 R_1)^{2i} (R_1 R_2 R_3)^2 (R_1 R_2)^{2i},$$

we have

$$A_1^2 A_2^2 \ldots A_q^2 = \{(R_2 R_1)^2 (R_1 R_2 R_3)^2\}^q = (R_2 R_3)^{2q} = E.$$

To see what this means geometrically, consider (for simplicity) the case $q = 3$. By numbering the angles of the hexagonal face of $\{6, 6\}$ as in Fig. 5.4a, we can describe the glide-reflections A_1, A_2, A_3 as taking the side 12 to 23, 34 to 45, 56 to 61, while their inverses do vice versa. In other words, the sides

$$12,\ 23,\ 34,\ 45,\ 56,\ 61$$

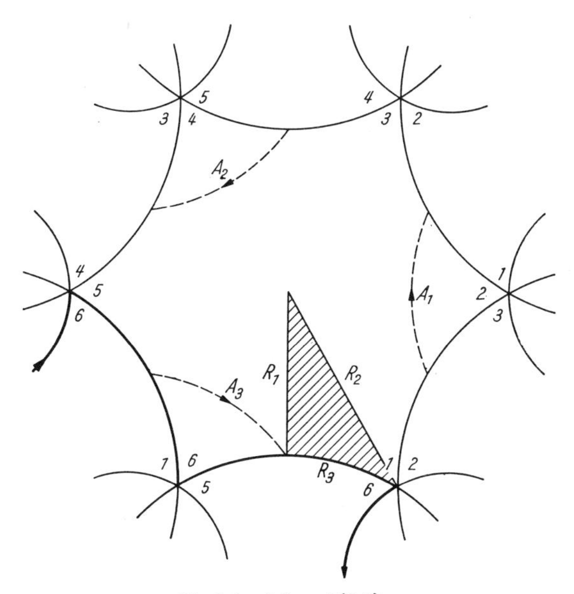

Fig. 5.4a. A face of $\{6, 6\}$

of the initial hexagon are the sides

$$23,\ 12,\ 45,\ 34,\ 61,\ 56$$

of its six neighbours. The six angles at any vertex of $\{6, 6\}$ are now numbered in their natural order (alternately clockwise and counter-clockwise); for we already see 123 at one vertex, 234 at the next, and so on. In the spirit of § 4.3, p. 36, the glide-reflection A_1 takes us out of the initial hexagon E by crossing the side 23, A_2 by crossing 45, and A_3 by crossing 61. Thus the expression

$$A_1^2 A_2^2 A_3^2 = A_1 A_1 A_2 A_2 A_3 A_3$$

represents a path crossing the side 61 of E, then side 61 of A_3, then the side 45 of A_3^2, then the side 45 of $A_2 A_3^2$, and so on; altogether, a circuit round the vertex 1 of E, as in Fig. 5.4b.

Since the same phenomenon arises in the general case, we deduce that the single relation

$$A_1^2 A_2^2 \ldots A_q^2 = E \tag{5.42}$$

(or 3.54) provides an abstract definition for the group generated by the q glide-reflections 5.41.

The Cayley diagram for 5.42, being the dual of $\{2q, 2q\}$, is another $\{2q, 2q\}$, such that the sides of each face are associated with the generators and directed as in the right half of Fig. 3.5c (where $q = 3$). For each generator, the corresponding edges form a set of disjoint Petrie polygons, together passing once through every vertex of $\{2q, 2q\}$ (see Fig. 4.5f for the case $q = 2$). Taking all the generators, we still have only half the Petrie polygons of $\{2q, 2q\}$, in agreement with our observation that 5.42 is not a *normal* subgroup of $[2q, 2q]$.

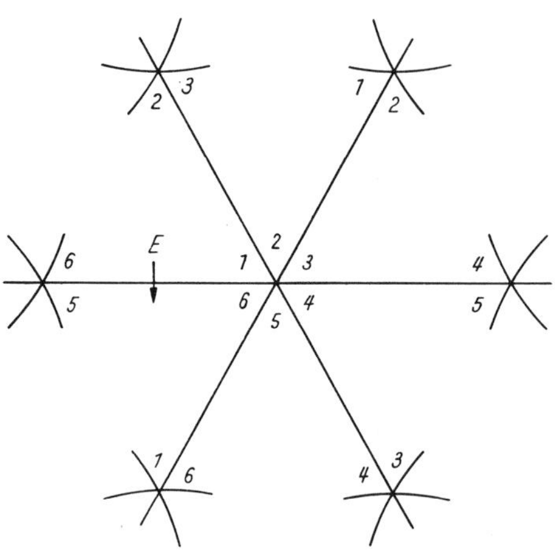

Fig. 5.4b. A vertex of $\{6, 6\}$

We proceed to make an abstract identification of all points that are related to one another by the group 5.42. This amounts to cutting out one face of the Cayley diagram and fastening together the pairs of edges marked alike (with directions agreeing). We thus obtain a closed non-orientable surface covered by a map having one vertex, q edges, and one face, so that its Euler-Poincaré characteristic is $2 - q$. In other words, as we remarked in § 3.5, 5.42 is the fundamental group for a closed non-orientable surface of the kind that can be regarded as a sphere with q cross-caps. (Note that our q is the k of THRELFALL 1932a, p. 33.)

When $q = 1$, we have $Z^2 = E$, the fundamental group for the projective plane. When $q = 2$, we have 3.53 or 4.504, the fundamental group for the Klein bottle (HILBERT and COHN-VOSSEN 1932, p. 308). The alternative relation $abab^{-1} = E$, corresponding to the identification of opposite sides of a rectangle, is derived from 4.504 by setting $P = ab$, $Q = b$.

5.5 The fundamental group for an orientable surface, obtained as a group of translations. We shall see that, when q is even, $[2q, 2q]$ has yet another subgroup whose fundamental region is a face of $\{2q, 2q\}$. When $q = 2$, this is simply the subgroup **p 1** of **p 4 m**, generated by translations along the sides of a face of $\{4, 4\}$.

Since q is now even, we write it as $2p$. Thus we are considering the group $[4p, 4p]$ defined by

$$R_1^2 = R_2^2 = R_3^2 = (R_1 R_2)^{4p} = (R_2 R_3)^{4p} = (R_3 R_1)^2 = E. \quad (5.51)$$

We can transform one face of $\{4p, 4p\}$ into its neighbours by *translations* along lines joining the mid-points of pairs of opposite sides. Since the polygon has $4p$ sides, these joining lines occur in orthogonal pairs, such as the lines of the reflections R_1 and

$$(R_2 R_1)^p R_1 (R_1 R_2)^p = R_2 (R_1 R_2)^{2p-1}$$

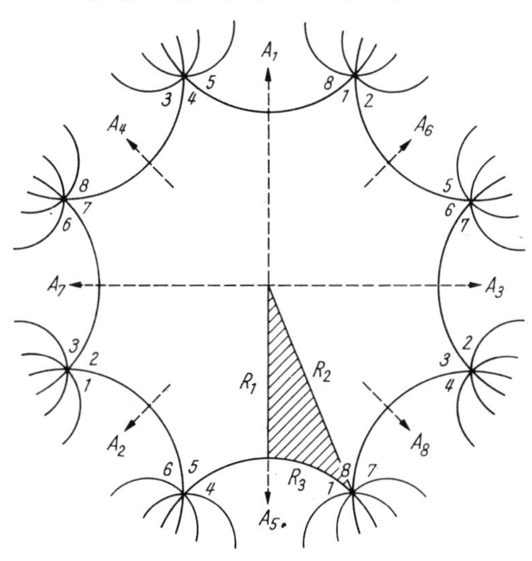

Fig. 5.5a. A face of {8, 8}

(see Fig. 5.5a for the case $p = 2$). This hyperbolic translation may be constructed either as the product of reflections in two lines, both orthogonal to its axis (COXETER 1957a, p. 202), or as the product of half-turns about two points on the axis. In the present case, the half-turns are $R_3 R_1$ and $(R_1 R_2)^{2p}$. Either method yields the translation

$$A_1 = R_3 R_2 (R_1 R_2)^{2p-1}, \quad (5.52)$$

which takes the "R_3" side to the opposite side.

This is one of the $2p$ generating translations, each along the join of the mid-points of a different pair of opposite sides of the $\{4p\}$. To obtain the others, we transform by powers of $R_1 R_2$. For a reason that will soon appear, we do not name them in the natural order of the vertices of the $\{4p\}$, but rather in the order of the vertices of the inscribed star-polygon

$$\left\{\frac{4p}{2p-1}\right\}$$

(COXETER 1963a, p. 93):

$$A_i = (R_2 R_1)^{(2p-1)(i-1)} A_1 (R_1 R_2)^{(2p-1)(i-1)} \tag{5.53}$$
$$= (R_2 R_1)^{(2p-1)(i-1)} R_3 R_2 (R_1 R_2)^{(2p-1)(i-1)}.$$

Of course, $A_{2p+1} = A_1^{-1}$, $A_{2p+2} = A_2^{-1}$, and so on. Moreover, since

$$A_{i+1} = (R_2 R_1)^{(2p-1)i} R_3 R_2 (R_1 R_2)^{(2p-1)(i+1)},$$

we have

$$A_1 A_2 \ldots A_{2p} A_1^{-1} A_2^{-1} \ldots A_{2p}^{-1} = A_1 A_2 \ldots A_{4p} = (R_3 R_2)^{4p} = E.$$

Our reason for numbering the A's according to the vertices of a star-polygon can be clarified by considering the case $p = 2$. Numbering the angles of an octagonal face of {8, 8} in the "octagrammatic" order 1 4 7 2 5 8 3 6, as in Fig. 5.5a, we can describe the translations $A_1, A_2, A_3, \ldots$ as taking the side 58 to 41, 61 to 52, 72 to 63, ... In other words, the side 41 of the initial octagon is the side 58 of the neighbouring octagon, and so on. By this device, the eight angles at any vertex of {8, 8} are numbered in their natural order; for we already see 123 at one vertex, 234 at another, etc. In the spirit of § 4.3, p. 36, the translation A_1 takes us out of the initial octagon E by crossing the side 14, A_2 by crossing 25, and so on. Thus the expression

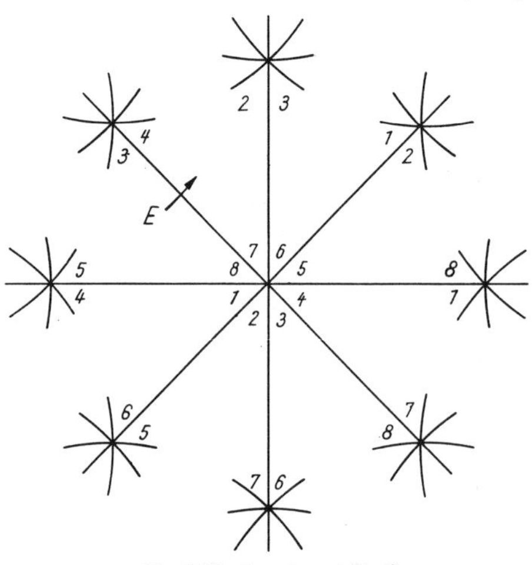

Fig. 5.5b. A vertex of {8, 8}

$$A_1 A_2 A_3 A_4 A_1^{-1} A_2^{-1} A_3^{-1} A_4^{-1} = A_1 A_2 A_3 A_4 A_5 A_6 A_7 A_8$$

represents a path crossing the side 83 of E, then the side 72 of A_8, then the side 61 of $A_7 A_8$, and so on; altogether, a circuit round the vertex 8 of E, as in Fig. 5.5b.

Since the same phenomenon arises for any p (where A_1 crosses the side 1 $2p$, and so on), we deduce that the single relation

$$A_1 A_2 \ldots A_{2p} A_1^{-1} A_2^{-1} \ldots A_{2p}^{-1} = E \tag{5.54}$$

(or $A_1 A_2 \ldots A_{2p} = A_{2p} \ldots A_2 A_1$) provides an abstract definition for the group generated by the $2p$ translations 5.53. Since, this time, our choice of generators was perfectly symmetrical, 5.54 is a *normal* subgroup (of index $8p$) in $[4p, 4p]$.

In terms of $R = R_1 R_2$ and $S = R_2 R_3$, we have

$$A_i = R^{-(2p-1)(i-1)} S^{-1} R^{(2p-1)i} \qquad (i = 1, 2, \ldots, 2p).$$

Thus 5.54 is also a normal subgroup (of index $4p$) in the rotation group $[4p, 4p]^+$, defined by

$$R^{4p} = S^{4p} = (RS)^2 = E$$

(NIELSEN 1932, p. 142).

The Cayley diagram for 5.54, being the dual of $\{4p, 4p\}$, is another $\{4p, 4p\}$, such that the sides of each face are associated with the generators and directed as in Fig. 3.6c (where $p = 2$).

We now make an abstract identification of all points that are related to one another by the group 5.54, i.e., we cut out one face of the Cayley diagram and fasten together the pairs of opposite edges (with directions agreeing). We thus obtain a closed orientable surface covered by a map having one vertex, $2p$ edges, and one face, so that its Euler-Poincaré characteristic is $2 - 2p$. In other words, as we remarked in § 3.5, 5.54 (or 3.52) is the fundamental group of a closed surface of genus p (THRELFALL 1932a, p. 36, Fig. 16).

When $p = 1$, we have $A_1 A_2 = A_2 A_1$, or $XY = YX$, the fundamental group for the torus. When $p = 2$, we have

$$A_1 A_2 A_3 A_4 A_1^{-1} A_2^{-1} A_3^{-1} A_4^{-1} = E,$$

the fundamental group for a "double torus", the surface of a solid figure-of-eight, beautifully illustrated by HILBERT and COHN-VOSSEN (1932, p. 228, Fig. 249) in terms of its less symmetrical definition

$$aba^{-1}b^{-1}cdc^{-1}d^{-1} = E$$

(THRELFALL 1932a, p. 1). In this other form, the same fundamental group is exhibited as a non-normal subgroup of $[8, 8]^+$. In fact, whenever $p > 1$, $[4p, 4p]^+$ has the normal subgroup 5.54 and several isomorphic subgroups that are not normal (THRELFALL 1932a, p. 37).

Chapter 6

The Symmetric, Alternating, and other Special Groups

The importance of the symmetric group is obvious from the fact that any finite group, of order n, say, is a subgroup of $\mathfrak{S}_n$. The procedure for obtaining generators and relations for $\mathfrak{S}_n$ can be carried over almost unchanged for a certain infinite group, first studied by ARTIN (1926); accordingly, we begin by describing this so-called braid group. The symmetric group (§ 6.2) has a subgroup $\mathfrak{A}_n$, of index 2 (§ 6.3), which is particularly interesting because, when $n > 4$, it is simple. In § 6.4, we exhibit the groups $\mathfrak{S}_3$, $\mathfrak{A}_4$, $\mathfrak{S}_4$ and $\mathfrak{A}_5$ as members of the family of polyhedral groups (l, m, n), defined by

$$R^l = S^m = T^n = RST = E.$$

In § 6.5 we consider the effect of omitting "$= E$" from these relations. Other generalizations of (l, m, n) are described in §§ 6.6 and 6.7. Finally, § 6.8 is a brief summary of the present state of knowledge on a problem due to BURNSIDE (1902) which only recently has been partially solved.

6.1 Artin's braid group. In Euclidean space, let $A_1B_1B_nA_n$ be a rectangle with n points evenly spaced along each of two opposite sides, say $A_1A_2 \ldots A_n$ and $B_1B_2 \ldots B_n$. Let the A's and the B's be joined by n simple non-intersecting curves μ_i (called "strings") running from A_i to B_{c_i} (where $c_1c_2 \ldots c_n$ is a permutation of $1\ 2 \ldots n$). Let ν_i be the orthogonal projection of μ_i on the plane of the rectangle. We call the arrangement of strings μ_i a *braid* if the curves ν_i cross one another in pairs a finite number of times and are monotonic, in the sense that each ν_i intersects every line between A_1A_n and B_1B_n and parallel to them in exactly one point.

We regard two braids as being equal if one can be continuously transformed into the other without breaking any strings (REIDEMEISTER 1932b, pp. 7—13, 41—43). The product of two braids is obtained by juxtaposing the two rectangles in such a way that the points $B_1B_2 \ldots B_n$ of the former coincide with the points $A_1A_2 \ldots A_n$ of the latter.

Calling A_1A_n a "horizontal" line, we may assume that the strings are vertical in general, but at certain levels two neighbouring strings interchange positions, one crossing in front of the other (BOHNENBLUST 1947, p. 127). The crossing of a string A_iB_{i+1} in front of a string $A_{i+1}B_i$ is denoted by σ_i. If the latter crosses in front of the former, the crossing is denoted by σ_i^{-1}. A braid is completely determined as a power product of the σ's. The braids $\sigma_i\sigma_k$ and $\sigma_k\sigma_i$, for $|i - k| > 1$, are easily seen to be equivalent (ARTIN 1926, p. 51), and the same holds for $\sigma_i\sigma_{i+1}\sigma_i$ and $\sigma_{i+1}\sigma_i\sigma_{i+1}$. In fact, the essentially distinct braids (formed by n strings) represent the elements of the *braid group*, which is the infinite group defined by

$$\sigma_i\sigma_{i+1}\sigma_i = \sigma_{i+1}\sigma_i\sigma_{i+1} \quad (1 \leqq i \leqq n-2), \qquad (6.11)$$

$$\sigma_i\sigma_k = \sigma_k\sigma_i \quad (i \leqq k-2) \qquad (6.12)$$

(ARTIN 1947a, b; BOHNENBLUST 1947; CHOW 1948; COXETER 1959a).

The relations 6.12, which say that σ_k and σ_i commute, are sometimes written in the abbreviated form

$$\sigma_i \rightleftarrows \sigma_k \qquad (6.13)$$

(NIELSEN 1924b, p. 169).

The braid group is actually generated by its two elements $\sigma = \sigma_1$ and

$$a = \sigma_1 \sigma_2 \ldots \sigma_{n-1}, \tag{6.14}$$

in terms of which

$$\sigma_i = a^{i-1} \sigma a^{-(i-1)} \tag{6.15}$$

(ARTIN 1926, p. 52). In terms of these two generators, the abstract definition is simply

$$a^n = (a\sigma)^{n-1}, \quad \sigma \rightleftarrows a^{-j} \sigma a^j \quad \left(2 \leqq j \leqq \frac{n}{2}\right) \tag{6.16}$$

(*ibid.*, p. 54).

CHOW (1948) observed that, when $n > 2$, the centre of the braid group is the $\mathfrak{C}_\infty$ generated by a^n. When $n = 2$, the braid group itself is the $\mathfrak{C}_\infty$ generated by a $(= \sigma = \sigma_1)$. When $n = 3$, the group

$$\sigma_1 \sigma_2 \sigma_1 = \sigma_2 \sigma_1 \sigma_2 \tag{6.17}$$

is generated by its two elements

$$a = \sigma_1 \sigma_2, \quad b = a\sigma = \sigma_1 \sigma_2 \sigma_1,$$

in terms of which $\sigma_1 = a^{-1} b$, $\sigma_2 = b a^{-1}$, and the abstract definition 6.16 reduces to

$$a^3 = b^2 \tag{6.18}$$

(ARTIN 1926, p. 70). Like 4.501 and 4.504, this group is defined by a single relation; for the theory of such groups, see SCHREIER (1924), MAGNUS (1932) and LYNDON (1950).

6.2 The symmetric group. The earliest presentations for the symmetric group $\mathfrak{S}_n$ were given by BURNSIDE (1897) and MOORE (1897). BURNSIDE's presentation

$$R^n = R_1^2 = (RR_1)^{n-1} = [R^{-r+1}(RR_1)^{r-1}]^r = (R^{-j} R_1 R^j R_1)^2 = E$$
$$\left(2 \leqq r \leqq n,\ 2 \leqq j \leqq \frac{n}{2}\right),$$

in terms of the generators

$$R = (1\ 2\ 3 \ldots n), \quad R_1 = (1\ 2),$$

has many redundant relations. In terms of the same generators, MOORE gave the simpler presentation

$$R^n = R_1^2 = (RR_1)^{n-1} = (R_1 R^{-1} R_1 R)^3 = (R_1 R^{-j} R_1 R^j)^2 = E \tag{6.21}$$
$$(2 \leqq j \leqq n-2).$$

MOORE also gave the presentation

$$\left.\begin{array}{ll} R_1^2 = R_2^2 = \cdots = R_{n-1}^2 = E, & \\ (R_i R_{i+1})^3 = E & (1 \leqq i \leqq n-2), \\ (R_i R_k)^2 = E & (i \leqq k-2), \end{array}\right\} \tag{6.22}$$

in terms of the generators

$$R_1 = (1\ 2),\ R_2 = (2\ 3), \ldots, R_{n-1} = (n-1\ n)$$

(BURNSIDE 1911, p. 464; CARMICHAEL 1923, p. 238). The two sets of generators are related by the equations

$$R_{i+1} = R^{-i} R_1 R^i \qquad (1 \leqq i \leqq n-2)$$

and

$$R = R_{n-1} R_{n-2} \ldots R_2 R_1 .$$

Obviously, 6.22 is equivalent to the three sets of relations

$$R_1^2 = E, \tag{6.23}$$

$$R_i R_{i+1} R_i = R_{i+1} R_i R_{i+1} \qquad (1 \leqq i \leqq n-2), \tag{6.24}$$

$$R_i R_k = R_k R_i \qquad (i \leqq k-2). \tag{6.25}$$

Since 6.24 and 6.25 define ARTIN's braid group[1]) (cf. 6.11 and 6.12), which is also defined by the relations

$$R^n = (R_1 R)^{n-1},\quad R_1 R^{-j} R_1 R^j = R^{-j} R_1 R^j R_1 \quad \left(2 \leqq j \leqq \frac{n}{2}\right) \tag{6.26}$$

(cf. 6.16), it follows that $\mathfrak{S}_n$ is defined by 6.23 and 6.26 (ARTIN 1926, p. 54; NIELSEN 1940) or by

$$R^n = (R_1 R)^{n-1},\quad R_1^2 = (R_1 R^{-j} R_1 R^j)^2 = E \quad \left(2 \leqq j \leqq \frac{n}{2}\right) \tag{6.27}$$

(cf. 6.21). COXETER (1937, p. 317) observed that, when n is even, we have the alternative presentation

$$R^n = (R_1 R)^{n-1},\quad R_1^2 = (R_1 R^{-1} R_1 R)^3 = (R_1 R^{-j} R_1 R^j)^2 = E \tag{6.271}$$

$$\left(2 \leqq j \leqq \frac{n}{2} - 1\right),$$

in which the relation $(R_1 R^{-1} R_1 R)^3 = E$ replaces $(R_1 R^{-n/2} R_1 R^{n/2})^2 = E$.

CARMICHAEL (1937, p. 169) preferred the presentation

$$S_i^2 = (S_i S_{i+1})^3 = (S_i S_{i+1} S_i S_j)^2 = E \tag{6.28}$$

$$(i, j = 1, 2, \ldots, n-1;\ j \neq i, i+1)$$

in terms of the transpositions

$$S_i = (i\ n) \qquad (i = 1, 2, \ldots, n-1)$$

(with $S_n = S_1$). In terms of

$$V_1 = S_1 = (1\ n),\quad V_j = S_1 S_j = (1\ j\ n) \quad (j = 2, 3, \ldots, n-1),$$

[1]) The braid group yields the symmetric group when we replace the strings μ_i by their projections ν_i, waiving the distinction between σ_i and σ_i^{-1}. Thus the R's obviously satisfy the same relations as the σ's and also 6.23 (which implies $R_i^2 = E$ since, by 6.24, all the R's are conjugate).

a still simpler presentation is

$$V_1^2 = V_j^3 = (V_i V_j)^2 = E \quad (1 \leqq i < j \leqq n-1) \qquad (6.281)$$

(COXETER 1934c, p. 218).

A geometrical interpretation for 6.22 is obtained by regarding $\mathfrak{S}_n$ as the group of permutations of the n vertices of a regular simplex $P_1 P_2 \ldots P_n$ in Euclidean $(n-1)$-space. These permutations are just the symmetry operations of the simplex. In particular, R_i is the reflection in the hyperplane that joins the mid-point of $P_i P_{i+1}$ to the remaining $n-2$ vertices. The $\binom{n}{2}$ edges $P_i P_j$ yield $\binom{n}{2}$ such hyperplanes, decomposing the circumsphere (or any concentric sphere) into $n!$ spherical simplexes. The R's appear as reflections in the bounding hyperplanes of one of these simplexes: the fundamental region (cf. Figs. 4.3, 4.5h, m, r, t).

In terms of n Cartesian coordinates, we may take P_i to be distant a from the origin along the ith axis, so that $P_1 P_2 \ldots P_n$ lies in the $(n-1)$-space $\sum x_i = a$. Then R_i, interchanging x_i and x_{i+1}, is the reflection in $x_i = x_{i+1}$, and the fundamental region is given by

$$x_1 \leqq x_2 \leqq \cdots \leqq x_n, \quad \sum x_i = a, \quad \sum x_i^2 = b \quad (b > a^2/n). \qquad (6.29)$$

The other spherical simplexes, being images of this one, are derived by permuting the x's in 6.29. The three parts of 6.22 express that the reflections are involutory, and that the angle between $x_i = x_{i+1}$ and $x_k = x_{k+1}$ is $\pi/3$ or $\pi/2$ according as $i = k-1$ or $i \leqq k-2$ (COXETER 1963a, pp. 80, 188).

The Cayley diagram is derived by taking a suitable point inside each spherical simplex (ROBINSON 1931). The simplest way to do this is to take the points whose coordinates are the permutations of $(0, 1, \ldots, n-1)$. In the fundamental region 6.29 we have $(0, 1, \ldots, n-1)$ itself. The edges of the diagram join these $n!$ points in pairs: each to its $n-1$ neighbours, distant $\sqrt{2}$. In fact, the diagram consists of the vertices and edges of a uniform polytope Π_{n-1}, whose two-dimensional faces are hexagons and squares representing the second and third parts of 6.22.

Trivially, Π_0 is a single point, Π_1 is a line-segment, and Π_2 is a regular hexagon, as in Fig. 3.3d (with the S-edges omitted) or Fig. 6.2 (where the fundamental region is one-sixth of the circle). Π_3 is the truncated octahedron, bounded by eight hexagons and six squares (KEPLER 1619, p. 125, Fig. 23[9]; KELVIN 1894, p. 15; STEINHAUS 1950, pp. 154—157). Π_4 is a four-dimensional polytope whose cells consist of ten truncated octahedra and twenty hexagonal prisms (HINTON 1906; pp. 135, 225; COXETER 1962c, p. 154).

According to FEDOROV (1885, pp. 286—298), a *parallelohedron* is a polyhedron that can be repeated by translations to fill the whole Euclidean (or affine) space. Thus any convex fundamental region for a 3-dimensional translation group is a parallelohedron. Using the same term in higher space, VORONOI gave a remarkably simple proof that an n-dimensional parallelohedron has at most $2(2^n - 1)$ bounding hyperplanes (VORONOI 1907, p. 107; 1908, p. 204; BAMBAH and DAVENPORT 1952, p. 225). This upper bound is attained by the polytope Π_n (SCHOUTE 1912), whose $(n-1)$-dimensional cells consist of $\binom{n+1}{i}$ generalized prisms $\Pi_{n-i} \times \Pi_{i-1}$ (COXETER 1963a, p. 124) for each value of i from 1 to n.

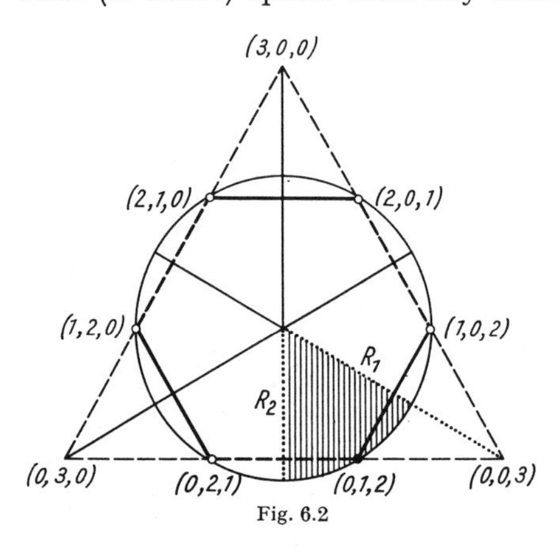

Fig. 6.2

Extending the notation of § 4.3, we may express 6.22 in the form

$$\mathfrak{S}_n \cong [3^{n-2}],$$

meaning [3, 3, . . . , 3] (TODD 1931, p. 225; COXETER 1963a, p. 199). Other groups generated by reflections will be discussed in Chapter 9.

6.3 The alternating group. MOORE (1897, p. 365) showed that, in terms of the generators

$$S_i = (1\ 2)(i+1\ i+2) \qquad (i = 1, 2, \ldots, n-2),$$

$\mathfrak{A}_n$ is defined by the relations

$$S_1^3 = S_j^2 = (S_{j-1} S_j)^3 = E \qquad (1 < j \leq n-2),$$

$$(S_i S_j)^2 = E \qquad (1 \leq i < j - 1,\ j \leq n-2).$$

CARMICHAEL (1923, p. 255; 1937, p. 172) gave the more symmetrical presentation

$$V_1^3 = \ldots = V_{n-2}^3 = (V_i V_j)^2 = E \qquad (1 \leq i < j \leq n-2)$$

in terms of the generators

$$V_i = (i\ n-1\ n) \qquad (i = 1, 2, \ldots, n-2),$$

which can evidently be replaced by $(1\ i+1\ n)$ (see also COXETER 1934c, p. 218).

When $n > 3$, $\mathfrak{A}_n$ has the generators

$$s = (3\ 4 \ldots n), \quad t = (1\ 2\ 3) \qquad (n \text{ odd})$$

or

$$s = (1\ 2)\ (3\ 4 \ldots n), \quad t = (1\ 2\ 3) \qquad (n \text{ even}).$$

Using them, CARMICHAEL (1923, p. 262) found the defining relations

$$s^{n-2} = t^3 = (st)^n = (ts^{-k}ts^k)^2 = E \quad \left(1 \leqq k \leqq \frac{n-3}{2}\right)$$

when n is odd, and

$$s^{n-2} = t^3 = (st)^{n-1} = [t^{(-1)^k} s^{-k} t s^k]^2 = E \quad \left(1 \leqq k \leqq \frac{n-2}{2}\right)$$

when n is even.

COXETER (1937b, pp. 317—318) reduced the number of relations to $[n/2] + 1$, namely, for $\mathfrak{A}_{2m+1}$,

$$R^{2m+1} = S^{2m+1} = (RS)^m, \ (R^{-j}S^j)^2 = E \quad (2 \leqq j \leqq m),$$

and for $\mathfrak{A}_{2m}$,

$$R^{2m-1} = S^{2m-1} = (RS)^m, \ (R^{-i}S^{-1}RS^i)^2 = E \quad (1 \leqq i \leqq m-1).$$

Extending the notation of § 4.4, p. 38 we write

$$\mathfrak{A}_n \cong [3^{n-2}]^+.$$

6.4 The polyhedral groups. The abstract definitions for $\mathfrak{S}_n$ and $\mathfrak{A}_n$ in §§ 6.2 and 6.3 are generally not the simplest ones when $n \leqq 6$ (see Table 5 on page 137). The groups $\mathfrak{A}_4$, $\mathfrak{S}_4$, and $\mathfrak{A}_5$, being isomorphic to the groups of rotations of the regular tetrahedron, octahedron and icosahedron, have long received special attention. In fact, one of the first abstract definitions appearing in the literature is the presentation

$$\iota^2 = \varkappa^3 = \lambda^5 = 1, \quad \lambda = \iota\varkappa$$

for $\mathfrak{A}_5$, given by HAMILTON (1856, p. 446), who was so pleased with it that he wrote: "I am disposed to give the name 'Icosian Calculus' to this system of symbols". DYCK (1882, p. 35) showed that $\mathfrak{A}_4$, $\mathfrak{S}_4$, and $\mathfrak{A}_5$ are the groups (2, 3, 3), (2, 3, 4) and (2, 3, 5) respectively, where (l, m, n) denotes

$$R^l = S^m = T^n = RST = E \qquad (6.41)$$

or

$$R^l = S^m = (RS)^n = E \qquad (6.42)$$

(BURNSIDE 1911, p. 408). In fact, as we saw in §§ 4.3, 4.4, and 5.3, the polyhedral group (l, m, n) is finite if and only if the number

$$k = lmn\left(\frac{1}{l} + \frac{1}{m} + \frac{1}{n} - 1\right) = mn + nl + lm - lmn \qquad (6.43)$$

is positive, i.e., in the cases

$$(2, 2, n), (2, 3, 3), (2, 3, 4), (2, 3, 5).$$

It is then generated by rotations through $2\pi/l$, $2\pi/m$, $2\pi/n$, about the three vertices of a spherical triangle whose angles are π/l, π/m, π/n. From the area of this triangle we deduce that the order of (l, m, n) is

$$2\left(\frac{1}{l}+\frac{1}{m}+\frac{1}{n}-1\right)^{-1}=\frac{2lmn}{k}. \qquad (6.44)$$

When $l = 2$ and $m = 3$, $k = 6 - n$; thus the order of $(2, 3, n) \cong [3, n]^+$ is

$$\frac{12n}{6-n} \qquad (1 < n \leqq 6).$$

One can easily remember the value of k: it is 1 for the icosahedral group (which is simple), 2 for the octahedral group (which has a subgroup of index 2), 3 for the tetrahedral group (which has a subgroup of index 3), and, finally, 4 for the dihedral group.

6.5 The binary polyhedral groups. THRELFALL (1932b) considered the larger group

$$\langle l, m, n\rangle \qquad (l, m, n > 1),$$

defined by

$$R^l = S^m = T^n = RST. \qquad (6.51)$$

Since (l, m, n) occurs as a factor group, $\langle l, m, n\rangle$ is infinite when $k \leq 0$.

We saw, in § 1.6, that when two of l, m, n are equal to 2, the relations

$$R^l = S^m = T^n = RST = Z \qquad (6.52)$$

imply

$$Z^2 = E. \qquad (6.521)$$

A similar argument (COXETER 1940c, p. 370) reveals the remarkable fact that the same holds for $\langle 2, 3, n\rangle$ ($n = 3, 4, 5$). Before we can assert that the period of Z is exactly 2, we must prove that the relations 6.52 do not imply $Z = E$. For this purpose we can either use the two-to-one correspondence between quaternions and rotations (COXETER 1946a, p. 139; 1962b, p. 87) or represent the generators by permutations of degree $8(n-2)$, as follows:

for $\langle 2, 3, 3\rangle$,

$R = (cac'a')(dbd'b')$,

$S = (bdab'd'a')(cc')$,

$T = (abca'b'c')(dd')$;

for $\langle 2, 3, 4\rangle$,

$R = (eae'a')(cbc'b')(dhd'h')(gfg'f')$,

$S = (cahc'a'h')(dged'g'e')(bb')(ff')$,

$T = (abcda'b'c'd')(efghe'f'g'h')$;

for $\langle 2, 3, 5\rangle$,

$$R = (aga'g')\,(dbd'b')\,(fcf'c')\,(kek'e')\,(jhj'h')\,(lil'i'),$$
$$S = (kdak'd'a')\,(cfbc'f'b')\,(jgej'g'e')\,(ilhi'l'h'),$$
$$T = (abcdea'b'c'd'e')\,(ghijkg'h'i'j'k')\,(ff')\,(ll').$$

In each case $Z = (aa')\,(bb')\ldots$, and the period of Z is indeed 2 whenever $k > 0$ (COXETER 1959b, pp. 66—67). On the other hand, the period of Z is infinite in all the three cases when $k = 0$. For, assuming that $l \le m \le n$, we may then derive a factor group by setting $R = T^{n/l}$ and $S = T^{n/m}$. Since $l^{-1} + m^{-1} + n^{-1} = 1$, all the relations 6.51 are automatically satisfied, and this factor group is the free group $\mathfrak{C}_\infty$ generated by T. Since the element $Z = T^n$ is of infinite period in this free group, it is still of infinite period in $\langle l, m, n\rangle$.

When $k > 0$, so that the period of Z is 2, we naturally call $\langle l, m, n\rangle$ a *binary* polyhedral group; its order is twice that of (l, m, n), namely

$$\frac{4lmn}{k}.$$

The particular cases are as follows: the dicyclic group $\langle 2, 2, n\rangle$, of order $4n$; the binary tetrahedral group $\langle 2, 3, 3\rangle$ (the "group of order 24 which does not contain a subgroup of order 12" of MILLER 1907, p. 4); the binary octahedral group $\langle 2, 3, 4\rangle$ (the "group of order 48 known as G_{52}" of MILLER 1907, p. 6); and the binary icosahedral group $\langle 2, 3, 5\rangle$ (the "compound perfect group of lowest possible order" of MILLER 1907, p. 10).

Setting $l = 2$ in 6.51, we obtain for $\langle 2, m, n\rangle$ the presentation

$$S^m = T^n = (ST)^2$$

or $TST = S^{m-1}$, $STS = T^{n-1}$. For instance, the binary tetrahedral group $\langle 2, 3, 3\rangle$ is $TST = S^2$, $STS = T^2$ or, in terms of S, T, $U = TS^{-1}$ and $V = T^{-1}S$,

$$TV = S,\ US = T,\ VT = U,\ SU = V.$$

In terms of T and $U = TS^{-1}$, the binary octahedral group $\langle 2, 3, 4\rangle$ has the presentation

$$TUT = UTU,\ TU^2T = U^2$$

(COXETER 1959b, pp. 79—80). In terms of T, $U = ST^{-1}$ and $V = T^{-1}S$, the binary icosahedral group $\langle 2, 3, 5\rangle$ has the presentation

$$UV^{-1}U^{-1}V = T,\ VT^{-1}V^{-1}T = U,\ TU^{-1}T^{-1}U = V$$

(ibid., p. 82). In terms of $A = VT^{-1}$ and $B = TU^{-1}$, this becomes

$$AB^2A = BAB,\ BA^2B = ABA$$

(ibid., p. 83).

One of the groups of order 24 listed in Table 1 is $\langle -2, 2, 3\rangle$. This is an instance of the group

$$\langle -l, m, n\rangle \qquad (l, m, n > 1)$$

defined by

$$R^{-l} = S^m = T^n = RST. \tag{6.53}$$

Writing $R^{-l} = S^m = T^n = RST = Z$ and

$$u = \frac{2mn}{k} - 1, \tag{6.54}$$

we find that, if k divides $2(m, n)$, where (m, n) denotes the greatest common divisor of m and n, the elements Z and

$$R_0 = R^u,\ S_0 = S^{-u},\ T_0 = T^{-u}$$

generate the same group in the form

$$R_0^l = S_0^m = T_0^n = R_0 S_0 T_0 = Z^{-u},\ Z \rightleftarrows R_0, S_0, T_0$$

(COXETER 1940c, pp. 373—374). Since u is odd, this is the direct product $\{R_0, S_0, T_0\} \times \{Z^2\}$, and we have

$$\langle -l, m, n\rangle \cong \langle l, m, n\rangle \times \mathfrak{C}_u,\quad k \mid 2(m, n), \tag{6.55}$$

of order

$$\frac{4lmnu}{k} = \frac{4lmn(mn - nl - lm + lmn)}{k^2}. \tag{6.56}$$

The particular cases are as follows:

$$\langle -2, 2, n\rangle \cong \langle 2, 2, n\rangle \times \mathfrak{C}_{n-1} \qquad (n \text{ even}),$$
$$\langle -2, 3, 3\rangle \cong \langle 2, 3, 3\rangle \times \mathfrak{C}_5,$$
$$\langle -4, 2, 3\rangle \cong \langle 2, 3, 4\rangle \times \mathfrak{C}_5 \qquad (\text{SEIFERT 1932, p. 8}),$$
$$\langle -3, 2, 4\rangle \cong \langle 2, 3, 4\rangle \times \mathfrak{C}_7,$$
$$\langle -2, 3, 4\rangle \cong \langle 2, 3, 4\rangle \times \mathfrak{C}_{11},$$
$$\langle -5, 2, 3\rangle \cong \langle 2, 3, 5\rangle \times \mathfrak{C}_{11} \qquad (\text{SEIFERT 1932, p. 8}),$$
$$\langle -3, 2, 5\rangle \cong \langle 2, 3, 5\rangle \times \mathfrak{C}_{19},$$
$$\langle -2, 3, 5\rangle \cong \langle 2, 3, 5\rangle \times \mathfrak{C}_{29}.$$

It is an interesting example of the principle of permanence that the order of $\langle -l, m, n\rangle$ is still given by 6.56 in the remaining cases, namely when k does not divide $2(m, n)$ (COXETER 1940c, pp. 375—378). Thus we have

$$\langle -2, 2, n\rangle \quad (u = n - 1), \text{ of order } 4n(n-1),$$
$$\langle -3, 2, 3\rangle \quad (u = 3), \text{ of order } 72 \quad (\text{SEIFERT 1932, p. 8}).$$

The above remarks suggest the idea of extending 6.52 by an element which commutes with R, S, T while its qth power is Z. Accordingly we define

$$\langle l, m, n \rangle_q$$

by the relations

$$R^l = S^m = T^n = RST = Z^q, \quad Z \rightleftarrows R, S, T. \tag{6.57}$$

(Thus $\langle l, m, n \rangle_1$ is $\langle l, m, n \rangle$ itself.) When $k > 0$, so that Z^q is of period 2, the order of $\langle l, m, n \rangle_q$ is q times that of $\langle l, m, n \rangle$, namely

$$4lmnq/k.$$

If $q = 2^c r$, where r is odd, we have the direct product of $\{R, S, T, Z^r\}$ and $\{Z^{2^{c+1}}\}$:

$$\langle l, m, n \rangle_{2^c r} \cong \langle l, m, n \rangle_{2^c} \times \mathfrak{C}_r. \tag{6.58}$$

In particular,

$$\langle l, m, n \rangle_r \cong \langle l, m, n \rangle \times \mathfrak{C}_r \qquad (r \text{ odd}), \tag{6.59}$$

as in the above treatment of $\langle -l, m, n \rangle$, where $r = u$.

Groups resembling $\langle l, m, n \rangle_q$ were described by THRELFALL and SEIFERT (1933, p. 577) as fundamental groups of fibred spaces.

6.6 Miller's generalization of the polyhedral groups. Another group having 6.42 for a factor group is

$$\langle l, m \mid n \rangle \qquad (l, m, n > 1),$$

defined by

$$R^l = S^m, \quad (RS)^n = E. \tag{6.61}$$

Its order is that of (l, m, n) multiplied by the period of the central element

$$Z = R^l = S^m.$$

If this period is finite, any divisor q yields a factor group

$$\langle l, m \mid n; q \rangle \cong \langle m, l \mid n; q \rangle$$

defined by

$$R^l = S^m = Z, \quad (RS)^n = Z^q = E. \tag{6.62}$$

In particular,

$$\langle l, m \mid n; 1 \rangle \cong (l, m, n). \tag{6.621}$$

Eliminating T from 6.52, we obtain $\langle l, m, n \rangle$ in the form

$$R^l = S^m = Z, \; (RS)^n = Z^{n-1}. \tag{6.622}$$

Hence, if n is odd,

$$\langle l, m \mid n; 2 \rangle \cong \langle l, m, n \rangle. \tag{6.623}$$

If $q = 2^c r$, where r is odd, we have

$$\langle l, m \mid n; q \rangle \cong \langle l, m \mid n; 2^c \rangle \times \mathfrak{C}_r. \tag{6.624}$$

In particular,

$$\langle l, m \mid n; r\rangle \cong (l, m, n)\times\mathfrak{C}_r \qquad (r \text{ odd}), \qquad (6.625)$$

and

$$\langle l, m \mid n;\ 2r\rangle \cong \langle l, m, n\rangle\times\mathfrak{C}_r \qquad (n \text{ and } r \text{ odd}). \quad (6.626)$$

In terms of S and $T = (RS)^{-1}$, $\langle l, m \mid n; 2\rangle$ is

$$S^m = (ST)^l = Z, \quad T^n = Z^2 = E.$$

If l is odd and n even, we can use the generators S and $T_1 = TZ$ so as to obtain

$$S^{2m} = T_1^n = (ST_1)^l = E, \quad S^m T_1 = T_1 S^m.$$

In particular, $\langle l, m \mid 2; 2\rangle$ with l odd is

$$S^{2m} = T^2 = (ST)^l = (S^m T)^2 = E. \qquad (6.627)$$

Defining k by 6.43 (p. 67) and writing

$$v = (l + m)\, n/k, \qquad (6.63)$$

we find that, if k divides $(l, m)\,(n - 1)$, the elements Z and

$$R_0 = RZ^{m(n-1)/k} = R^v, \quad S_0 = SZ^{l(n-1)/k} = S^v,$$
$$T_0 = (RS)^{-1}\, Z^{(l+m)/k} = (RS)^{-1}\, Z^{v/n}$$

generate $\langle l, m \mid n\rangle$ in the form

$$R_0^l = S_0^m = T_0^n = R_0 S_0 T_0 = Z^v, \ Z \rightleftarrows R_0, S_0, T_0.$$

Thus, in the notation of 6.57, if k divides $(l, m)\,(n - 1)$,

$$\langle l, m \mid n\rangle \cong \langle l, m, n\rangle_v, \qquad (6.631)$$

of order

$$\frac{4lmnv}{k} = \frac{4(l + m)\, lmn^2}{k^2}.$$

Inserting the extra relation $Z^v = E$, we obtain the factor group

$$\langle l, m \mid n; v\rangle \cong (l, m, n)\times\mathfrak{C}_v, \quad k \mid (l, m)\,(n - 1). \quad (6.632)$$

If v is odd, 6.59 yields the stronger result

$$\langle l, m \mid n\rangle \cong \langle l, m, n\rangle\times\mathfrak{C}_v, \quad k \mid (l, m)\,(n - 1). \quad (6.633)$$

The particular cases are as follows:

$$\langle 2, 2 \mid n\rangle \cong \langle 2, 2, n\rangle\times\mathfrak{C}_n \qquad (n \text{ odd}),$$
$$\langle 2, 4 \mid 3\rangle \cong \langle 2, 3, 4\rangle\times\mathfrak{C}_9 \qquad (\text{MILLER 1907, p. 9}),$$
$$\langle 2, 5 \mid 3\rangle \cong \langle 2, 3, 5\rangle\times\mathfrak{C}_{21} \qquad (\textit{ibid.}, \text{p. 11}),$$
$$\langle 2, 3 \mid 5\rangle \cong \langle 2, 3, 5\rangle\times\mathfrak{C}_{25} \qquad (\textit{ibid.}, \text{p. 13}).$$

We have also

$$\langle 3, 3 \mid 2 \rangle \cong \langle 2, 3, 3 \rangle_4, \text{ of order } 96 \qquad \text{(MILLER 1907, p. 3)},$$

$$\langle 3, 5 \mid 2 \rangle \cong \langle 2, 3, 5 \rangle_{16}, \text{ of order } 1920 \qquad \text{(\textit{ibid.}, p. 12)},$$

and their factor groups

$$\langle 3, 3 \mid 2; 2^c \rangle \cong \mathfrak{A}_4 \times \mathfrak{C}_{2^c} \qquad (c = 0, 1, 2),$$

$$\langle 3, 5 \mid 2; 2^c \rangle \cong \mathfrak{A}_5 \times \mathfrak{C}_{2^c} \qquad (c = 0, 1, 2, 3, 4).$$

In the case $l = 2$, $m = 3$, $n = 3$, k does not divide $(l, m)(n - 3)$; nevertheless MILLER (1907, p. 4) found that

$$\langle 2, 3 \mid 3 \rangle \cong \langle 2, 3, 3 \rangle \times \mathfrak{C}_5.$$

It is remarkable that Z still has period $2v$ in the remaining cases:

$$\langle 2, 2 \mid n \rangle \cong \langle 2, 2 \mid n; 2n \rangle \qquad \text{(MILLER 1909, p. 168)},$$

$$\langle 2, m \mid 2 \rangle \cong \langle 2, m \mid 2; m + 2 \rangle,$$

$$\langle 3, 4 \mid 2 \rangle \cong \langle 3, 4 \mid 2; 14 \rangle \cong \langle 3, 4 \mid 2; 2 \rangle \times \mathfrak{C}_7 \qquad \text{(MILLER 1907, p. 9)},$$

$$\langle 2, 3 \mid 4 \rangle \cong \langle 2, 3 \mid 4; 20 \rangle \cong \langle 2, 3 \mid 4; 4 \rangle \times \mathfrak{C}_5 \qquad \text{(\textit{ibid.}, p. 8)}.$$

In fact, $\langle 2, m \mid 2 \rangle$ is 1.85 with $n = 2$ and $r = -m - 1$; therefore its order is $2(r^2 - 1) = 2m(m + 2)$. In particular, by 6.625,

$$\langle 2, m \mid 2 \rangle \cong \mathfrak{D}_m \times \mathfrak{C}_{m+2} \qquad (m \text{ odd}).$$

The group

$$\langle -l, m \mid n \rangle \qquad (l, m, n > 1),$$

defined by

$$R^{-l} = S^m, \quad (RS)^n = E, \tag{6.64}$$

or $R^l S^m = (RS)^n = E$, or

$$S^m = (ST)^l, \quad T^n = E, \tag{6.641}$$

is clearly infinite when $l = m$, since the extra relation $T = E$ yields a factor group $\mathfrak{C}_\infty$. In particular, $\langle -2, 2 \mid n \rangle$ has a factor group

$$\langle -2, 2 \mid n; q \rangle$$

of order $2nq$, which was discussed by MILLER (1908, p. 7; our n and q are his k and $n/2$). Clearly, also,

$$\langle -l, m \mid n \rangle \cong \langle -m, l \mid n \rangle.$$

Thus we may restrict ourselves to the cases where $l < m$.

Defining k by 6.43, and writing

$$w = (m - l)\, n/k, \tag{6.65}$$

we find that, if k divides (l, m), the elements $Z = R^{-l} = S^m$ and

$$R_0 = RZ^{(m+2n-mn)/k} = R^w, \quad S_0 = SZ^{(ln-l-2n)/k} = S^{-w},$$
$$T_0 = (RS)^{-1} Z^{-(m-l)/k} = TZ^{-w/n}$$

generate $\langle -l, m \mid n \rangle$ in the form

$$R_0^l = S_0^m = T_0^n = R_0 S_0 T_0 = Z^{-w}, \quad Z \rightleftarrows R_0, S_0, T_0.$$

Thus, if k divides (l, m),

$$\langle -l, m \mid n \rangle \cong \langle l, m, n \rangle_w, \tag{6.651}$$

of order

$$\frac{4lmnw}{k} = \frac{4(m-l)\,lmn^2}{k^2}.$$

Also

$$\langle -l, m \mid n; w \rangle \cong (l, m, n) \times \mathfrak{C}_w, \quad k \mid (l, m), \tag{6.652}$$

and if w is odd,

$$\langle -l, m \mid n \rangle \cong \langle l, m, n \rangle \times \mathfrak{C}_w, \quad k \mid (l, m). \tag{6.653}$$

The particular cases are as follows:

$$\langle -2, 4 \mid 3 \rangle \cong \langle 2, 3, 4 \rangle \times \mathfrak{C}_3 \quad \text{(MILLER 1909, p. 172)},$$
$$\langle -2, 3 \mid 5 \rangle \cong \langle 2, 3, 5 \rangle \times \mathfrak{C}_5 \quad (\textit{ibid.}, \text{p. } 179),$$
$$\langle -2, 5 \mid 3 \rangle \cong \langle 2, 3, 5 \rangle \times \mathfrak{C}_9 \quad (\textit{ibid.}, \text{p. } 174).$$

We have also

$$\langle -3, 5 \mid 2 \rangle \cong \langle 2, 3, 5 \rangle_4, \text{ of order } 480.$$

We proceed to prove that Z still has period $2w$ when k does not divide (l, m), namely in the cases:

$$\langle -2, m \mid 2 \rangle \cong \langle -2, m \mid 2; m-2 \rangle, \text{ of order } 2m(m-2),$$
$$\langle -2, 3 \mid 3 \rangle \cong \langle -2, 3 \mid 3; 2 \rangle \cong \langle 2, 3 \mid 3; 2 \rangle \cong \langle 2, 3, 3 \rangle,$$
$$\langle -2, 3 \mid 4 \rangle \cong \langle -2, 3 \mid 4; 4 \rangle, \text{ of order } 96,$$
$$\langle -3, 4 \mid 2 \rangle \cong \langle -3, 4 \mid 2; 2 \rangle \cong \langle 3, 4 \mid 2; 2 \rangle.$$

In fact, $\langle -2, m \mid 2 \rangle$ is 1.85 with $n = 2$, $r = m - 1$; therefore its order is $2(r^2 - 1) = 2m(m-2)$. In particular, by 6.625,

$$\langle -2, m \mid 2 \rangle \cong \mathfrak{D}_m \times \mathfrak{C}_{m-2} \quad (m \text{ odd}).$$

Moreover, $\langle -2, 3 \mid 3 \rangle$ is defined by the relations

$$R^{-2} = S^3 = Z, \quad (RS)^3 = E,$$

which imply $S^{-1}R^{-1} = S^2 R$ and

$$R^{-1}Z^2 R = R^{-1}S^6 R = (R^{-1}S^2 R)^3 = (R^{-1}S^{-1}R^{-1})^3 = (SRS)^3$$
$$= S^{-1}(S^2 R)^3 S = S^{-1}(S^{-1}R^{-1})^3 S = E,$$

i.e., $Z^2 = E$. Hence another definition for $\langle -2, 3 \mid 3\rangle$ is

$$R^2 = S^3 = Z,\ (RS)^3 = Z^2 = E,$$

and it follows from 6.623 that

$$\langle -2, 3 \mid 3\rangle \cong \langle 2, 3, 3\rangle \tag{6.66}$$

(MILLER 1909, p. 172).

Furthermore, $\langle -2, 3 \mid 4\rangle$ is defined by the relations

$$R^{-2} = S^3 = Z,\ (RS)^4 = E$$

which imply $S^{-1}R^{-1} = S^2R$, $RS = R^{-1}SZ^{-1}$ and

$$Z^4 = Z^4(SR)^4 = (ZSR)^4 = (S^4R)^4.$$

Transforming this last relation by S^2R, we obtain

$$\begin{aligned} Z^4 &= R^{-1}S^{-2}(S^4R)^4S^2R = (R^{-1}S^2RS^2R)^4 = (R^{-1}S^{-1}R^{-1}S^{-1}R^{-1})^4 \\ &= (SRS)^4 = S^{-1}(S^2R)^4S = S^{-1}(S^{-1}R^{-1})^4S = E. \end{aligned}$$

Hence, an equivalent set of defining relations is

$$R^{-2} = S^3 = Z,\ (R^{-1}S)^4 = Z^4 = E.$$

Replacing R by R^{-1} we find that

$$\langle -2, 3 \mid 4\rangle \cong \langle 2, 3 \mid 4; 4\rangle. \tag{6.67}$$

Finally, $\langle -3, 4 \mid 2\rangle$ may be derived from $\langle 2, 3, 3\rangle$, in the form

$$R^3 = S^3 = (RS)^2$$

(COXETER 1940c, p. 367), by adjoining an element T of period 2 which transforms R into S. The extended group, of order 48, is defined by

$$S^3 = (ST)^4,\ T^2 = E,$$

i.e., it is $\langle -4, 3 \mid 2\rangle$ in the form 6.641. Writing $R_1 = (ST)^{-1}$, we see (from 6.521) that the relations

$$R_1^{-4} = S^3 = Z,\ (R_1S)^2 = E$$

imply $Z^2 = E$. Replacing R_1 by R^{-1}, we deduce

$$\langle -3, 4 \mid 2\rangle \cong \langle 3, 4 \mid 2; 2\rangle. \tag{6.671}$$

It follows from 6.627 that $\langle -3, 4 \mid 2\rangle$ is the group

$$T^2 = (ST)^3 = (S^4T)^2 = E, \tag{6.672}$$

of order 48, considered by BURNSIDE (1911, pp. 401, 419).

Comparing 6.67 and 6.671 with our previous description of $\langle 3, 4 \mid 2\rangle$ and $\langle 2, 3 \mid 4\rangle$, we obtain

$$\langle 3, 4 \mid 2\rangle \cong \langle -3, 4 \mid 2\rangle \times \mathfrak{C}_7 \tag{6.68}$$

and

$$\langle 2, 3 \mid 4 \rangle \cong \langle -2, 3 \mid 4 \rangle \times \mathfrak{C}_5. \qquad (6.681)$$

It is interesting to note that the group $\langle -2, 3 \mid n \rangle$, namely

$$R^{-2} = S^3, \ (RS)^n = E,$$

is generated by its elements

$$U = RS, \ V = SR$$

(in terms of which $R = UVU$, $S = U^{-1}V^{-1}$) in the pleasing form

$$U^n = E, \ UVU = VUV. \qquad (6.69)$$

In this way we exhibit $\langle -2, 3 \mid n \rangle$ as a factor group of the braid group 6.17, which is $\langle -2, 3 \mid \infty \rangle$.

6.7 A new generalization. Another group having 6.42 for a factor group is

$$l\,[2n]\,m \qquad (l, m, n > 1),$$

defined by

$$R^l = S^m = E, \quad (RS)^n = (SR)^n. \qquad (6.71)$$

Its order is that of (l, m, n) multiplied by the period of the central element

$$Z = (RS)^n = (SR)^n.$$

Clearly $l\,[2n]\,m \cong m\,[2n]\,l$, and

$$2\,[2n]\,2 \cong \mathfrak{D}_{2n}. \qquad (6.72)$$

Still defining k by 6.43, we find that, if k divides (l, m), the elements Z and

$$R_0 = RZ^{-m/k}, \quad S_0 = SZ^{-l/k}, \quad T_0 = (RS)^{-1} Z^{(l+m-lm)/k}$$

generate $l\,[2n]\,m$ in the form

$$R_0^l = S_0^m = T_0^n = R_0 S_0 T_0 = Z^{-lm/k}, \ Z \rightleftarrows R_0, S_0, T_0.$$

Thus

$$l\,[2n]\,m \cong \langle l, m, n \rangle_{lm/k}, \qquad k \mid (l, m), \quad (6.73)$$

in the notation of 6.57; the period of Z is $2lm/k$, and the order of the group is

$$\frac{2lm}{k}\,\frac{2lmn}{k} = \frac{4l^2m^2n}{k^2}.$$

Using 6.59 and 6.58, we have:

$$3\,[4]\,3 \cong \langle 2, 3, 3\rangle_3 \cong \langle 2, 3, 3\rangle \times \mathfrak{C}_3,$$

$$3\,[4]\,5 \cong \langle 2, 3, 5\rangle_{15} \cong \langle 2, 3, 5\rangle \times \mathfrak{C}_{15},$$

$$2\,[6]\,4 \cong \langle 2, 3, 4\rangle_5, \text{ of order } 192,$$

$$2\,[10]\,3 \cong \langle 2, 3, 5\rangle_6 \cong \langle 2, 3, 5\rangle_2 \times \mathfrak{C}_3,$$

$$2\,[6]\,5 \cong \langle 2, 3, 5\rangle_{10} \cong \langle 2, 3, 5\rangle_2 \times \mathfrak{C}_5,$$

where $\langle 2, 3, 5\rangle_2$ is the group of order 240 defined by

$$R^2 = S^3 = T^5 = RST = Z^2,\ Z \rightleftarrows R, S, T.$$

It is remarkable that the period of Z is still $2lm/k$ in the remaining cases:

$$2\,[4]\,m, \text{ of order } 2m^2,$$

$$2\,[6]\,3, \text{ of order } 48,$$

$$2\,[8]\,3, \text{ of order } 144,$$

$$3\,[4]\,4, \text{ of order } 288.$$

In fact, the relations

$$R^2 = S^m = E,\ (RS)^2 = (SR)^2 = Z,$$

defining $2\,[4]\,m$, imply $RSR \rightleftarrows S$ and

$$Z^m = (RSRS)^m = (RSR)^m S^m = RS^m RS^m = E$$

(cf. SHEPHARD 1952, p. 93). We see from 4.5051 that $2\,[4]\,m$ is a factor group of

$$\mathbf{cm} \cong 2\,[4]\,\infty.$$

When m is odd, $2\,[4]\,m$ is generated by R and S^2, and so also by R, $S^{-2}Z$ and Z. Since the period of $S^{-2}Z$ is m, while that of $S^{-2}ZR = S^{-1}RS$ is 2, the two elements R and $S^{-2}Z$ generate the dihedral group $\mathfrak{D}_m$. Since Z itself generates the cyclic group $\mathfrak{C}_m$, we have

$$2\,[4]\,m \cong \mathfrak{D}_m \times \mathfrak{C}_m \qquad (m \text{ odd}). \qquad (6.74)$$

From $\langle -2, 3 \mid m\rangle$, in the form

$$S^m = E,\ STS = TST \qquad (6.75)$$

(cf. 6.69), we may derive $2\,[6]\,m$ as follows. The element

$$Q = STS = TST$$

of 6.75 transforms the generators S and T into

$$Q^{-1}SQ = T,\ Q^{-1}TQ = S.$$

Adjoining a new element R, of period 2, which transforms $\langle -2, 3 \mid m \rangle$ according to this inner automorphism, we obtain the group

$$S^m = E, STS = TST, R^2 = E, RSR = T. \qquad (6.751)$$

Eliminating T, we find that 6.751 is in fact $2[6]\,m$:

$$R^2 = S^m = E, \ (RS)^3 = (SR)^3 \qquad (m = 3, 4, 5).$$

In particular, the order of $2[6]\,3$ is thus seen to be twice that of

$$\langle -2, 3 \mid 3 \rangle \cong \langle 2, 3, 3 \rangle.$$

In $2[8]\,3$, the elements R and $S_1 = SZ^{-1}$ satisfy

$$R^2 = E, S_1^3 = (RS_1)^4 = Z^{-3}.$$

But the relations

$$R^2 = E, S_1^3 = (RS_1)^4 = Z_1$$

define $\langle -3, 4 \mid 2 \rangle$, and imply $Z_1^2 = E$. Hence $2[8]\,3$ is the direct product of $\{R, S_1\}$ and $\{Z^2\}$:

$$2[8]\,3 \cong \langle -3, 4 \mid 2 \rangle \times \mathfrak{C}_3. \qquad (6.76)$$

Finally, we observe that in $3[4]\,4$, the elements $R_1 = RZ$ and S satisfy

$$S^4 = E, R_1^3 = (R_1S)^2 = Z^3.$$

But the relations

$$S^4 = E, R_1^3 = (R_1S)^2 = Z_1$$

define $\langle -2, 3 \mid 4 \rangle$ and imply $Z_1^4 = E$. Hence $3[4]\,4$ is the direct product of $\{R_1, S\}$ and $\{Z^4\}$:

$$3[4]\,4 \cong \langle -2, 3 \mid 4 \rangle \times \mathfrak{C}_3. \qquad (6.77)$$

When q is odd, the dihedral group $\mathfrak{D}_q$ may be expressed as

$$S^2 = E, \ RSR \ldots SR = SRS \ldots RS,$$

where the last relation, having q factors on each side, may be regarded as $(RS)^n = (SR)^n$ with the fractional value $q/2$ for n. With this interpretation, 6.72, in the form $2[q]\,2 \cong \mathfrak{D}_q$, continues to hold when q is odd. In the same spirit, $m[3]\,m$ is an alternative symbol for $\langle -2, 3 \mid m \rangle$ in the form 6.75, while $3[5]\,3$ is

$$S^3 = E, \ RSRSR = SRSRS$$

(which is easily seen to be $\langle 2, 3, 5 \rangle_3$, whereas $\langle -2, 5 \mid 3 \rangle \cong \langle 2, 3, 5 \rangle_9$). Since the element $Q = RS \ldots R = SR \ldots S$ transforms R into S, and S into R, the relations

$$S^m = E, RS \ldots R = SR \ldots S \quad (q \text{ factors on each side}) \qquad (6.78)$$

always imply $R^m = E$, and thus suffice to define

$$m\,[q]\,m \qquad (q \text{ odd}).$$

By adjoining an involutory element that transforms 6.78 in the same manner as Q (cf. 6.751), we obtain $2\,[2q]\,m$. Hence the order of $l\,[q]\,m$ is

$$g = \frac{8\,l^2 m^2 q}{\{(l+m)\,q - lm\,(q-2)\}^2}$$

not only when q is even but also when q is odd. In the latter case we must have $l = m$, since R and S are conjugate. To sum up,

There is a group $l\,[q]\,m$, *defined by* 6.71 (*with* $n = q/2$) *or* 6.78, *whenever the integers* l, m, q *satisfy*

$$l > 1,\ m > 1,\ q > 2,\ (l+m)\,q > lm\,(q-2),$$

with the restriction that, when q *is odd,* $l = m$. *The order is* $g = 2\,h^2/q$ *where*

$$h = \frac{2\,lmq}{(l+m)\,q - lm\,(q-2)},\quad \frac{2}{h} = \frac{1}{l} + \frac{1}{m} + \frac{2}{q} - 1.$$

These groups are important because of their occurrence in the theory of regular complex polygons (SHEPHARD 1952, p. 92). In the complex affine plane with a unitary metric, a *reflection* is a congruent transformation leaving invariant all the points on a line; its period may be greater than 2. *A regular complex polygon* is a finite connected configuration of points (called vertices) and lines (called edges), invariant under two unitary reflections: one, say R, which cyclically permutes the vertices on one edge, and another, say S, which cyclically permutes the edges through one of these vertices. It follows that the group $\{R, S\}$, of order g, say, is transitive on the vertices and on the edges. The polygon is said to be of type

$$l\,(g)\,m$$

if R is of period l, and S of period m, so that there are l vertices on each edge and m edges through each vertex. Since there is a vertex for each coset of $\{S\}$ and an edge for each coset of $\{R\}$, there are altogether g/m vertices and g/l edges (SHEPHARD 1952, p. 88).

In view of the discovery (COXETER 1962b) that every finite group $l\,[q]\,m$ is the symmetry group of a pair of reciprocal complex polygons

$$l\,(g)\,m,\quad m\,(g)\,l,$$

it is clearly desirable to replace these symbols by

$$l\{q\}\,m,\quad m\{q\}\,l\,.$$

Here q may be described as the number of vertices in a minimal cycle such that every two consecutive vertices, but no three, belong to an edge.

The above inequalities enable us to make the complete list of such regular complex polygons:

$$2\{q\}2 = \{q\},\quad m\{4\}2,\quad 2\{4\}m,\quad 3\{3\}3,\quad 3\{6\}2,\quad 2\{6\}3,\quad 3\{4\}3,$$

$$4\{3\}4,\quad 3\{8\}2,\quad 2\{8\}3,\quad 4\{6\}2,\quad 2\{6\}4,\quad 4\{4\}3,\quad 3\{4\}4,\quad 3\{5\}3,$$

$$5\{3\}5,\quad 3\{10\}2,\quad 2\{10\}3,\quad 5\{6\}2,\quad 2\{6\}5,\quad 5\{4\}3,\quad 3\{4\}5.$$

This notation extends in a natural manner to polytopes in more than two (complex) dimensions:

$$3\{3\}3\{4\}2,\quad 2\{4\}3\{3\}3,\quad 3\{3\}3\{3\}3,\quad 3\{3\}3\{3\}3\{3\}3,$$
$$m\{4\}2\{3\}2\ldots\{3\}2,\quad 2\{3\}2\ldots\{3\}2\{4\}m.$$

For a full account see *Regular Complex Polytopes* (Coxeter 1974).

6.8 Burnside's problem. A group all of whose elements satisfy the relation

$$X^n = E$$

for some fixed integer n (so that the periods of all the elements divide n) is said to be of *exponent* n. Burnside (1902) brought into prominence the following problem: Is a group necessarily finite if it has m generators and is of exponent n? Accordingly, we define a maximal group $B_{m,n}$, maximal in the sense that every group having these properties is a factor group of $B_{m,n}$. Burnside conjectured that $B_{m,n}$ is finite for every m and n.

When $n = 2$ the problem is trivial; for $B_{m,2}$, defined by

$$S_i^2 = (S_i S_j)^2 = E \qquad (i, j = 1, 2, \ldots, m),$$

is $\mathfrak{C}_2^m$, the Abelian group of order 2^m and type $(1, 1, \ldots, 1)$. Burnside (1902, pp. 231—233) showed that $B_{m,3}$ is finite (for another proof see Schenkman 1954), and gave wrongly 3^{2^m-1} as its order; Levi and van der Waerden (1933) showed that the order is $3^{f(m)}$ where

$$f(m) = \binom{m}{1} + \binom{m}{2} + \binom{m}{3} = \frac{m(m^2+5)}{6}. \tag{6.81}$$

Of course $B_{1,3} \cong \mathfrak{C}_3$; and

$$B_{2,3} \cong (3, 3 \mid 3, 3)$$

is one of the groups of order 27 given in Table 1 (Coxeter 1939, p. 74). Leech (1963, p. 264) has found that $B_{3,3}$ is given by assigning period 3 to the elements

$$A,\ B,\ C,\ BC,\ CA,\ AB,\ B^{-1}C,\ C^{-1}A,\ A^{-1}B,$$
$$ABC,\ A^{-1}BC,\ AB^{-1}C,\ ABC^{-1}.$$

BURNSIDE (1902, pp. 234—237) showed that $B_{2,4}$ is finite, and gave 2^{12} as its order. Actually, he proved only that the order *divides* 2^{12}. Indeed, PHILIP HALL once believed it to be 2^{10} (NEUMANN 1937b, p. 506, footnote). But BURNSIDE's value 2^{12} has been checked another way by S. TOBIN (1960). LEECH (1963, p. 264) has found that $B_{2,4}$ is given by assigning period 4 to the elements

$$A,\ B,\ A\,B,\ A^{-1}B,\ A\,B^2,\ A^2B$$

and to any three of

$$A^{-1}B^{-1}A\,B, \quad A^2B^2, \quad A^{-1}B\,A\,B, \quad A\,B^{-1}A\,B\,.$$

(For other groups resembling these, see COXETER 1940d, p. 249 [1]).)

The groups $B_{m,4}$ were shown to be finite by SANOV (1940); a description of the proof is given by G. HIGMAN (1957, pp. 123—128). The order of $B_{m,4}$ is not known for $m > 3$, although some bounds have been determined. It is rumoured that the order of $B_{3,4}$ is 2^{69}.

M. HALL (1958; 1959, p. 337) proved that $B_{m,6}$ is finite for every m. Earlier, P. HALL and HIGMAN (1956, p. 38) showed that if $B_{m,6}$ is finite, then its order is $2^a 3^{f(b)}$ where

$$a = 1 + (m - 1)\ 3^{f(m)}, \quad b = 1 + (m - 1)\ 2^m,$$

and $f(m)$ is given by 6.81.

So far, $B_{m,2}$, $B_{m,3}$, $B_{m,4}$ and $B_{m,6}$ are the only cases in which $B_{m,n}$ is known to be finite. BURNSIDE (1902, pp. 237—238) showed that, for a prime p, the order of $B_{2,p}$ (if finite) is not less than p^{2p-3}. In particular, the order of $B_{2,5}$ is not less than 5^7. MAGNUS (1950, p. 126) improved this and showed that the order of $B_{2,5}$ is at least 5^{14}.

P. S. NOVIKOV and S. I. ADJAN (1968) have shown that for $m \geqq 2$ and any odd $n \geqq 665$, $B_{m,n}$ is infinite.

BAER (1944) has given a full account of the connection between BURNSIDE's problem and the general theory of groups. If $\mathfrak{S}$ and $\mathfrak{T}$ are subsets of a group $\mathfrak{G}$, $(\mathfrak{S}, \mathfrak{T})$ denotes the subgroup generated by the commutators $(S, T) = S^{-1}T^{-1}S\,T$. The *lower central series*

$${}^0\mathfrak{G} = \mathfrak{G}, \quad {}^{i+1}\mathfrak{G} = (\mathfrak{G}, {}^i\mathfrak{G})$$

is said to be *finite* if ${}^{c-1}\mathfrak{G}$ is a proper subgroup of ${}^c\mathfrak{G}$, and ${}^c\mathfrak{G} = {}^{c+1}\mathfrak{G} = \cdots$. If, further, the series terminates with $\{E\} = {}^c\mathfrak{G}$, then $\mathfrak{G}$ is said to be *nilpotent of class c* (ZASSENHAUS 1958, pp. 141, 142). BAER proved that,

[1]) In that paper, on p. 248, just before (5), $4n^4$ is a misprint for $8n^4$.

if n is a power of a prime, the following conditions are necessary and sufficient for $B_{m,n}$ to be finite:

a) E is the only element common to every ${}^iB_{m,n}$;

b) the lower central series of $B_{m,n}$ is finite.

Of course, if $B_{m,n}$ satisfies a) and b), it is nilpotent. BAER (1944, p. 153) pointed out that a) may be replaced by the weaker condition:

a*) If $B_{m,n}/{}^iB_{m,n} \neq \{E\}$, then ${}^{i+1}B_{m,n}$ is a proper subgroup of ${}^iB_{m,n}$.

Conditions b) and a*) have been investigated by GRÜN (1936), MAGNUS (1935a; 1937), WITT (1937), ZASSENHAUS (1940), and LYNDON (1954).

Even when $B_{m,n}$ itself is infinite, it is possible that the orders of the finite m-generator groups of exponent n are bounded. This is the "restricted Burnside conjecture". If it is true for some m and n, then there is a maximal finite m-generator group, $B^*_{m,n}$, of which all others are factor groups. If n is a prime power, then the restricted Burnside conjecture is true if and only if $B_{m,n}$ satisfies condition b). In fact, in this case

$$B^*_{m,n} \cong B_{m,n}/\mathfrak{N},$$

where $\mathfrak{N}$ is the normal subgroup of $B_{m,n}$ consisting of those elements common to every ${}^iB_{m,n}$ ($i = 1, 2, \ldots$) (BAER 1944, p. 153). GRÜN (1940) asserted that condition b) is satisfied for $m = 2$, $n = 5$; an error in his proof was found by SANOV (1951). The correctness of the assertion was proved by KOSTRIKIN (1955), who showed that the order of $B^*_{2,5}$ is not greater than 5^{34}. HIGMAN (1956) proved that the restricted Burnside conjecture is true for every m when $n = 5$.

A group is called a *p-group* if the periods of all the elements are powers of the prime p. A finite p-group is simply a group whose order is a power of p. Every finite p-group is nilpotent (P. HALL 1933). Since every factor group of a nilpotent group is nilpotent, it follows that if the restricted Burnside conjecture is true for some m and some prime power n, then the class of a finite m-generator group of exponent n is not greater than the class of $B^*_{m,n}$. Thus it is interesting to construct finite m-generator groups with given exponent n and of large nilpotent class. GREEN (1952) constructed a finite 2-generator group, of exponent $n = p^\alpha \geqq 5$ with p an odd prime, which has class $c = 2n - 2$. MEIER-WUNDERLI (1956, p. 174) increased this to $c \geqq 2n - 1$ for the case when $\alpha = 1$.

Another question raised by BURNSIDE (1902, p. 230) is whether a group with a given finite number of generators is necessarily finite if the periods of its elements do not exceed a given number h. NEUMANN (1937a) has solved this "bounded periods" problem for $h = 3$, and SANOV (1940) for $h = 4$.

Chapter 7

Modular and Linear Fractional Groups

Groups of linear transformations may be constructed over all kinds of fields, finite or infinite, discrete or continuous (DICKSON 1901a, b; DIEUDONNÉ 1955). In this chapter we begin by describing the n-dimensional unimodular group $\mathfrak{M}_n$, which is the group of automorphisms of the Abelian group $\mathfrak{F}_1^n$ or $\mathfrak{C}_\infty^n$. For the special case when $n = 2$, we find the abstract definition

$$(RU)^2 = (R^2U^2)^6 = (R^3U^2)^2 = E$$

(7.29): MAGNUS (1935b, p. 367) proved that when $n > 3$, $\mathfrak{M}_n$ is the factor group obtained by adding the single extra relation 7.33 to the relations defining Γ_n, the group of automorphisms of the free group $\mathfrak{F}_n$; NIELSEN (1924a) had proved it earlier for $n = 3$. After describing DICKSON's linear fractional group $LF(n, q)$, we compare various presentations for $LF(2, p)$, where p is prime, and for the related groups $SL(2, p)$, $GL(2, p)$. SINKOV's definition for $LF(2, 2^m)$ is given in § 7.6. Finally, we describe new definitions for the simple groups of orders 5616 (§ 7.7), 7920 and 95040 (§ 7.8).

7.1 Lattices and modular groups. All n-dimensional lattices (as used in crystallography or the geometry of numbers) are affinely equivalent. In other words, real affine n-space contains essentially just one lattice. We may take the lattice points to be the points

$$X = (x_1, x_2, \ldots, x_n)$$

with integral coordinates, and identify them with their position vectors. The lattice is generated by the unit vectors

$$S_1 = (1, 0, 0, \ldots, 0),\ S_2 = (0, 1, 0, \ldots, 0), \ldots,\ S_n = (0, 0, 0, \ldots, 1).$$

An affine collineation which leaves the lattice as a whole invariant, and keeps the origin

$$(0, 0, 0, \ldots, 0)$$

fixed, is determined by its effect on the unit vectors. If it takes S_i into the lattice point

$$S_i' = (a_{i1}, a_{i2}, \ldots, a_{in}),$$

it may be described by the matrix

$$A = \begin{pmatrix} a_{11} & a_{21} & \cdots & a_{n1} \\ a_{12} & a_{22} & \cdots & a_{n2} \\ \cdots & & & \cdots \\ a_{1n} & a_{2n} & \cdots & a_{nn} \end{pmatrix}$$

i.e., it takes $X = (x_1, x_2, \ldots, x_n)$ into

$$X' = (x_1', x_2', \ldots, x_n') = (x_1, x_2, \ldots, x_n) \begin{pmatrix} a_{11} \cdots a_{n1} \\ \cdots \\ a_{1n} \cdots a_{nn} \end{pmatrix} = XA .$$

This choice of A, instead of its transpose, makes our decision to read products from left to right agree with the classical rule for multiplying matrices (VAN DER WAERDEN 1931, p. 112). If the collineation is equi-affine (i.e., if it preserves content), then the vectors S_i' also generate the lattice and

$$\begin{vmatrix} a_{11} \cdots a_{n1} \\ \cdots \\ a_{1n} \cdots a_{nn} \end{vmatrix} = \pm 1.$$

Hence the group of all affine collineations which leave the lattice invariant and keep a point fixed is the *unimodular* group $\mathfrak{M}_n$: the group of all $n \times n$ matrices of integers of determinant ± 1 (HUA and REINER 1951, p. 331).

$\mathfrak{M}_n$ is a subgroup of the larger group which consists of those affine collineations which leave the lattice invariant. In the larger group, the lattice translations form a normal subgroup whose quotient group is isomorphic to $\mathfrak{M}_n$. The group $\mathfrak{M}_n$ has a centre of order 2, generated by the *central inversion*

$$Z = \begin{pmatrix} -1 & 0 \cdots & 0 \\ 0 & -1 \cdots & 0 \\ \cdots & \cdots & \\ 0 & 0 \cdots & -1 \end{pmatrix},$$

which reverses the sign of every coordinate. The central quotient group $\mathfrak{M}_n/\{Z\}$ is the *projective unimodular* group $\mathfrak{P}_n$: the group of "rational" collineations in real projective $(n-1)$-space (HUA and REINER 1952, p. 467).

The unimodular group $\mathfrak{M}_n$ has a subgroup of index 2, the *modular* group $\mathfrak{M}_n^+$, consisting of matrices of determinant 1. When n is odd, any matrix of determinant -1 is the result of multiplying a matrix of determinant 1 by Z; hence

$$\mathfrak{M}_n^+ \cong \mathfrak{P}_n, \; \mathfrak{M}_n \cong \mathfrak{P}_n \times \{Z\} \qquad (n \text{ odd}).$$

On the other hand, when n is even, so that the matrix Z has determinant 1, the sense-preserving collineations in $\mathfrak{P}_n$ form a subgroup of index 2, the *projective modular* group $\mathfrak{P}_n^+$, which is the central quotient group of $\mathfrak{M}_n^+$:

$$\mathfrak{P}_n^+ \cong \mathfrak{M}_n^+/\{Z\}.$$

In particular, $\mathfrak{P}_2^+$ is the group of linear fractional transformations

$$z' = \frac{az + b}{cz + d}$$

with integral coefficients satisfying $ad - bc = 1$.

7.2 Defining relations when $n = 2$. The unimodular group $\mathfrak{M}_n$ is generated by the matrices

$$U_1 = \begin{pmatrix} 0\,1\,0 \cdots 0\,0 \\ 0\,0\,1 \cdots 0\,0 \\ 0\,0\,0 \cdots 0\,0 \\ \cdots \\ 0\,0\,0 \cdots 0\,1 \\ 1\,0\,0 \cdots 0\,0 \end{pmatrix}, \quad U_2 = \begin{pmatrix} 1\,0\,0 \cdots 0\,0 \\ 1\,1\,0 \cdots 0\,0 \\ 0\,0\,1 \cdots 0\,0 \\ \cdots \\ 0\,0\,0 \cdots 1\,0 \\ 0\,0\,0 \cdots 0\,1 \end{pmatrix},$$

$$U_3 = \begin{pmatrix} -1\,0\,0 \cdots 0\,0 \\ 0\,1\,0 \cdots 0\,0 \\ 0\,0\,1 \cdots 0\,0 \\ \cdots \\ 0\,0\,0 \cdots 1\,0 \\ 0\,0\,0 \cdots 0\,1 \end{pmatrix}, \quad U_4 = \begin{pmatrix} 0\,1\,0 \cdots 0\,0 \\ 1\,0\,0 \cdots 0\,0 \\ 0\,0\,1 \cdots 0\,0 \\ \cdots \\ 0\,0\,0 \cdots 1\,0 \\ 0\,0\,0 \cdots 0\,1 \end{pmatrix}$$

(MacDuffee 1933, p. 34). Hua and Reiner (1949, p. 421) showed that U_1, U_2, U_3 suffice. In fact,

$$U_4 = U^* U_2^{-1} U^* U_3,$$

where

$$U^* = U_1 U_2 V U_1^{-1} V^{-1} U_1 V U_2^{-1} V^{-1} U_1^{-1}, \; V = (U_1 U_2)^{n-2}.$$

Trott (1962) showed that U_1 and U_2 suffice to generate M_n when n is even (they generate M_n^+ for odd n); when n is odd, M_n is generated by U_2 and $U_1^* = U_1 U_3$.

When $n = 2$, it is convenient to rename U_1, U_2, U_3 as R_1, $R_3 R_2$, R_3, so that $\mathfrak{M}_2$ is generated by

$$R_1 = \begin{pmatrix} 0 & 1 \\ 1 & 0 \end{pmatrix}, \quad R_2 = \begin{pmatrix} -1 & 0 \\ 1 & 1 \end{pmatrix}, \quad R_3 = \begin{pmatrix} -1 & 0 \\ 0 & 1 \end{pmatrix}$$

of determinant -1, which satisfy the relations

$$R_1^2 = R_2^2 = R_3^2 = E, \; (R_1 R_2)^3 = (R_1 R_3)^2 = Z, \; Z^2 = E. \quad \mathbf{(7.21)}$$

Any element of $\mathfrak{M}_2$ has determinant 1 or -1 according as it is the product of an even or odd number of R's; therefore the subgroup $\mathfrak{M}_2^+$, of index 2, is generated by the two matrices

$$S = R_1 R_2 = \begin{pmatrix} 1 & 1 \\ -1 & 0 \end{pmatrix}, \quad T = R_1 R_3 = \begin{pmatrix} 0 & 1 \\ -1 & 0 \end{pmatrix},$$

of determinant 1, which satisfy

$$S^3 = T^2 = Z, \ Z^2 = E. \tag{7.22}$$

HUA and REINER (1951, p. 331) showed that those elements of $\mathfrak{M}_2^+$ which involve T an even number of times form a subgroup $\mathfrak{N}_2$ of index 2, which is the commutator subgroup of $\mathfrak{M}_2$ (cf. FRASCH 1933, p. 245, footnote). Thus $\mathfrak{N}_2$ is generated by S and

$$W = T^{-1}ST = \begin{pmatrix} 0 & -1 \\ 1 & 0 \end{pmatrix} \begin{pmatrix} -1 & 1 \\ 0 & -1 \end{pmatrix} = \begin{pmatrix} 0 & 1 \\ -1 & 1 \end{pmatrix},$$

which satisfy

$$S^3 = W^3 = Z, \ Z^2 = E. \tag{7.23}$$

The corresponding linear fractional transformations (for which it is convenient to use the same symbols) generate $\mathfrak{P}_2$, $\mathfrak{P}_2^+$, and the commutator subgroup of $\mathfrak{P}_2$. They satisfy the same relations with $Z = E$, namely:

$$R_1^2 = R_2^2 = R_3^2 = (R_1 R_2)^3 = (R_1 R_3)^2 = E, \tag{7.24}$$

$$S^3 = T^2 = E, \tag{7.25}$$

$$S^3 = W^3 = E. \tag{7.26}$$

In order to show that the above six sets of relations suffice to define the respective abstract groups, we consider the case of $\mathfrak{P}_2$ and 7.24. (The remaining five follow easily from this one.) Interpreting the complex number $z = x + yi$ as the point (x, y) in the Euclidean (or strictly, conformal) plane, we observe that a circle with its centre on the x-axis has an equation of the form

$$A(x^2 + y^2) + 2Bx + C = 0 \text{ or } Az\bar{z} + B(z + \bar{z}) + C = 0$$

or

$$\left(z + \frac{B}{A}\right)\left(\bar{z} + \frac{B}{A}\right) = \frac{B^2 - AC}{A^2},$$

where A, B, C are real numbers. The centre and radius are $-B/A$ and $\sqrt{B^2 - AC}/A$. Inversion in this circle interchanges the points z and z', where

$$\left(z' + \frac{B}{A}\right)\left(\bar{z} + \frac{B}{A}\right) = \frac{B^2 - AC}{A^2},$$

so that

$$Az'\bar{z} + B(z' + \bar{z}) + C = 0$$

and

$$z' = -\frac{B\bar{z} + C}{A\bar{z} + B} \tag{7.27}$$

(FORD 1929, p. 11). Since $B^2 - AC > 0$, we can adjust the coefficients so that

$$B^2 - AC = 1.$$

In the special (or limiting) case when $A = 0$ (and, say, $B = -1$), we have the expression

$$z' = -\bar{z} + C \tag{7.271}$$

for the reflection in the line $z + \bar{z} = C$ or $x = C/2$. Any linear fractional transformation

$$z' = \frac{az + b}{cz + d}, \tag{7.28}$$

with real coefficients satisfying $ad - bc = 1$, can be expressed as the product of such an inversion and reflection (or of two reflections). In fact, the relations

$$z' = -\frac{d\bar{z} + (d^2 - 1)/c}{c\bar{z} + d}, \quad z'' = -\bar{z}' + \frac{a-d}{c} \qquad (c \neq 0)$$

imply

$$z'' = \frac{az + (ad - 1)/c}{cz + d},$$

while the relations $z' = -\bar{z}$ and $z'' = -\bar{z}' + b$ imply $z'' = z + b$.

The transformations 7.27 and 7.271 are naturally associated with the involutory matrices

$$\begin{pmatrix} -B & A \\ -C & B \end{pmatrix}, \quad \begin{pmatrix} -1 & 0 \\ C & 1 \end{pmatrix}$$

of determinant -1. In particular, the three transformations

$$z' = 1/\bar{z}, \ z' = -\bar{z} + 1, \ z' = -\bar{z},$$

associated with the generators

$$\begin{pmatrix} 0 & 1 \\ 1 & 0 \end{pmatrix}, \quad \begin{pmatrix} -1 & 0 \\ 1 & 1 \end{pmatrix}, \quad \begin{pmatrix} -1 & 0 \\ 0 & 1 \end{pmatrix}$$

of $\mathfrak{P}_2$, are represented by inversion in $x^2 + y^2 = 1$, reflection in $x = 1/2$, and reflection in $x = 0$.

Poincaré (1882, p. 8) regarded the half-plane $y > 0$ as a conformal representation of the hyperbolic plane. The lines of the hyperbolic plane are represented by half-lines and semicircles orthogonal to the x-axis. The above transformations can thus be interpreted in the hyperbolic plane as reflections in the sides of a triangle with angles 0, $\pi/2$, $\pi/3$ (Klein 1879a, p. 121; Fricke and Klein 1897, p. 432). It follows that the group $\mathfrak{P}_2$ is $[3, \infty]$ in the notation of 4.32, and 7.24 is a complete abstract definition for it.

Clearly, 7.25 is the subgroup generated by $S = R_1R_2$ and $T = R_1R_3$, and 7.26 is the subgroup generated by S and $W = TST$.

From 7.26 we can derive 7.25 by adjoining an involutory element T which transforms S into W. From 7.25 we can derive 7.24 by adjoining an involutory element R_1 which transforms S into S^{-1} and T into

T^{-1} $(= T)$ and then writing

$$R_2 = R_1 S,\ R_3 = R_1 T.$$

Combining two of the right-angled triangles, we obtain an isosceles triangle with angles 0, $\pi/3$, $\pi/3$, which serves as a fundamental region for the projective modular group $\mathfrak{P}_2^+ \cong [3,\infty]^+$. The arrangement of such triangles, filling the upper half-plane, has been nicely drawn by KLEIN (1879a, p. 120). To obtain a fundamental region for the commutator subgroup we could combine two of them to form a "rhomb" with angles 0, $2\pi/3$, 0, $2\pi/3$.

For a purely algebraic treatment, see REIDEMEISTER (1932a, pp. 44 to 46).

The above definitions for $\mathfrak{M}_2$ and $\mathfrak{P}_2$, namely 7.21 and 7.24, involve three generators. But we can use instead the two generators

$$R = R_1 R_2 R_3,\ U_2 = R_3 R_2$$

(COXETER and TODD 1936, p. 195), in terms of which $\mathfrak{M}_2$ is defined by the relations

$$(R U_2)^2 = (R^3 U_2^2)^2 = (R^2 U_2^2)^6 = E, \tag{7.29}$$

and $\mathfrak{P}_2$ by

$$(R U_2)^2 = (R^3 U_2^2)^2 = (R^2 U_2^2)^3 = E. \tag{7.291}$$

By expressing 7.24 in the form

$$I^3 = J^2 = K^2 = (KI)^2 = (KJ)^2 = E,$$

we see that $\mathfrak{P}_2$ is the group of automorphisms of the simplest braid group, 6.17 or 6.18 (SCHREIER 1924, p. 169).

7.3 Defining relations when $n \geqq 3$. Any n-dimensional lattice is a Cayley diagram for $\mathfrak{C}_\infty^n$, the direct product of n infinite cyclic groups. The automorphisms of $\mathfrak{C}_\infty^n$ correspond to the equiaffine collineations which leave the lattice invariant and keep one point fixed, namely that one which represents the identity element. Hence $\mathfrak{M}_n$ is the group of automorphisms of $\mathfrak{C}_\infty^n$ (NIELSEN 1924a; JACOBSON 1943, p. 4).

The elements of $\mathfrak{C}_\infty^n$ are the lattice points, and the group operation is defined by

$$XY = (x_1 + y_1, x_2 + y_2, \ldots, x_n + y_n).$$

In terms of the generators $S_1, S_2, \ldots, S_n$ (§ 7.1, p. 83), an abstract definition of $\mathfrak{C}_\infty^n$ is

$$S_i S_j = S_j S_i \qquad (i, j = 1, 2, \ldots, n).$$

An automorphism of $\mathfrak{C}_\infty^n$ is defined by its effect on these generators. The transformations

$$P = U_4,\quad Q = U_1,\quad O = U_3,\quad U = U_4 U_2 U_4$$

(§ 7.2) generate $\mathfrak{M}_n$. Since

$$S_1 P = S_2,\ S_2 P = S_1,\ S_3 P = S_3, \ldots, S_n P = S_n,$$

we may write, in the notation of NIELSEN (1924b, p. 171),

$$P = [S_2, S_1, S_3, \ldots, S_n].$$

Similarly

$$Q = [S_2, S_3, \ldots, S_n, S_1],\ O = [S_1^{-1}, S_2, \ldots, S_n],\ U = [S_1 S_2, S_2, \ldots, S_n].$$

These transformations P, Q, O, U are also automorphisms of the free group $\mathfrak{F}_n$, whose n generators we denote by the same symbols $S_1, S_2, \ldots, S_n$. In fact, NIELSEN (1924b, pp. 169—209) showed that P, Q, O, U generate Γ_n, the group of automorphisms of $\mathfrak{F}_n$. NEUMANN (1932, pp. 368—369, 374; NEUMANN and NEUMANN 1951, § 6) slightly simplified NIELSEN's defining relations so as to obtain, when $n \geqq 3$ (in the notation of 6.13):

a) $P^2 = E,$ (7.31)

b) $(QP)^{n-1} = Q^n,$

c) $P \rightleftarrows Q^{-i} P Q^i \quad \left(2 \leqq i \leqq \frac{n}{2}\right),$

d) $O^2 = E,$

e) $O \rightleftarrows Q^{-1} P Q,$

f) $O \rightleftarrows QP,$

g) $(PO)^4 = E,$

h) $U \rightleftarrows Q^{-2} P Q^2 \quad (n > 3),$

i) $U \rightleftarrows Q P Q^{-1} P Q,$

j) $U \rightleftarrows Q^{-2} O Q^2,$

k) $U \rightleftarrows Q^{-2} U Q^2 \quad (n > 3),$

l) $U \rightleftarrows OUO,$

m) $U \rightleftarrows P Q^{-1} O U O Q P,$

n) $U \rightleftarrows P Q^{-1} P Q P U P Q^{-1} P Q P,$

o) $(P Q^{-1} U Q)^2 = U Q^{-1} U Q U^{-1},$

p) $U^{-1} P U P O U O P O = E,$

q) $(P O P U)^2 = E.$

But q) is redundant, since a), d) and p) imply $U^{-1} P U P O U = O P O$, whence $(P U P O)^2 = E$.

Clearly, $\{P, Q\}$ is the symmetric group $\mathfrak{S}_n$, and we recognize 7.31 a)—c) as its definition 6.27. These relations therefore imply

$$(Q P)^{n-1} = Q^n = E.$$

Similarly $\{P, Q, O\}$, defined by 7.31 a)—g), is a group of order $2^n n!$ (the Ω_n of NIELSEN 1924b, p. 172) which may be regarded as the symmetry group of the n-dimensional Cartesian frame. ALFRED YOUNG (1930) named it the *hyper-octahedral* group. Ω_n can be generated by two generators (NEUMANN 1932, pp. 370—373; COXETER and TODD 1936, p. 199), as can Γ_n (NEUMANN 1932, pp. 375—378).

Any automorphism of $\mathfrak{F}_n$ induces an automorphism of its factor group $\mathfrak{C}_\infty^n$. By mapping the generators P, Q, O, U of Γ_n onto the generators P, Q, O, U of $\mathfrak{M}_n$, we see that every automorphism of $\mathfrak{C}_\infty^n$ is induced by an automorphism of $\mathfrak{F}_n$. This homomorphism of Γ_n upon $\mathfrak{M}_n$ is an isomorphism if and only if $n = 1$. For instance, if $n > 1$, we have

$$[S_1 S_2, S_2, \ldots, S_n] = [S_2 S_1, S_2, \ldots, S_n] \tag{7.32}$$

in $\mathfrak{M}_n$, but not in Γ_n.

What must be added to the relations 7.31 (which define Γ_n) to yield the factor group $\mathfrak{M}_n$? Certainly we will add the relation that expresses 7.32, namely $U = O U^{-1} O$ or

$$(O U)^2 = E. \tag{7.33}$$

In fact no further relations are needed. We will verify this for $n = 2$ and $n = 3$; it had been proved for $n = 3$ by NIELSEN (1924a, p. 24) and for $n > 3$ by MAGNUS (1935b, p. 367).

It follows that the factor group $\mathfrak{P}_n \cong \mathfrak{M}_n/\{Z\}$ is defined by 7.31 and 7.33 along with the further relation

$$(O Q)^n = E, \tag{7.34}$$

which expresses the fact that, in $\mathfrak{P}_n$,

$$[S_1^{-1}, S_2^{-1}, \ldots, S_n^{-1}] = [S_1, S_2, \ldots, S_n] = E.$$

The derivation of $\mathfrak{M}_n$ from Γ_n by adding the single relation 7.33 can already be observed when $n = 2$, in which case the relations 7.31 are replaced by

$$P^2 = O^2 = (P O)^4 = (P O P U)^2 = (P O U)^3 = E, \quad (O U)^2 = (U O)^2$$

(NEUMANN 1932, p. 374). By substituting

$$P = R_1, \quad O = R_3, \quad U = R_1 R_3 R_1 R_2$$

in 7.21, we obtain $\mathfrak{M}_2$ in the form

$$P^2 = O^2 = (P O)^4 = (P O P U)^2 = (P O U)^3 = (O U)^2 = E.$$

In terms of

$$R_1 = P, \; R_2 = POPU, \; R_3 = O,$$

we can express Γ_2 in the form

$$R_1^2 = R_2^2 = R_3^2 = E, \; (R_1 R_3)^2 = Z, \; Z^2 = (R_1 R_2 Z)^3 = E,$$
$$(R_2 Z)^2 \rightleftarrows R_3,$$

which yields 7.21 when we add $(R_2 Z)^2 = E$.

We now verify the next case, $n = 3$, by showing that $\mathfrak{M}_3$ is defined by 7.31 (with $n = 3$) and 7.33.

First we observe that the relations 7.31 a)—p) and 7.33 are equivalent to 7.31 a)—k), m)—o) and

$$O = PUPU^{-1}PU. \qquad (7.35)$$

In fact, 7.35 is a simple consequence of 7.31 d), p) and 7.33. Conversely, from 7.31 a), d) and 7.35, we have

$$OU^{-1} = PUPU^{-1}P, \text{ or } (UO)^{-1} = PUP(PU)^{-1},$$

which implies 7.33; the latter relation makes 7.31 l) "empty" and enables us to derive p) from 7.35.

When $n = 3$, the relations 7.31 a)—k), m)—o) and 7.35 reduce as follows. Since $P^2 = (QP)^2 = E$, o) yields

$$(QPUQ)^2 = UQ^{-1}UQU^{-1}.$$

Since $Q^3 = E$, j) yields $U \rightleftarrows QOQ^{-1}$. Since a), d) and 7.35 imply 7.33, m) may be written as

$$U \rightleftarrows PQ^{-1}U^{-1}QP, \text{ or } (UQP)^2 = (QPU)^2, \text{ or } U \rightleftarrows QPUQP.$$

Since a) and b) imply $PQ^{-1}PQP = PQ$, n) may be written as $U \rightleftarrows PQUPQ$. Thus the defining relations for Γ_3, with 7.33 added, imply

$$P^2 = (QP)^2 = Q^3 = O^2 = E, \; O = PUPU^{-1}PU,$$
$$(QPUQ)^2 = UQ^{-1}UQU^{-1}, \qquad (7.36)$$
$$O \rightleftarrows QP, \; U \rightleftarrows QOQ^{-1}, \; U \rightleftarrows QPUQP, \; U \rightleftarrows PQUPQ.$$

To establish 7.36 as a sufficient definition for $\mathfrak{M}_3$, we observe that it can be derived from a presentation due to NIELSEN (1924a, p. 24) by omitting his relations

$$(OU)^2 = (POPU)^2 = E$$

which follow, as we have seen, from $P^2 = O^2 = E$, $O = PUPU^{-1}PU$,

Since $\mathfrak{M}_3 \cong \mathfrak{P}_3 \times \{Z\}$, where

$$Z = [S_1^{-1}, S_2^{-1}, S_3^{-1}] = (OQ)^3,$$

we can derive $\mathfrak{P}_3 \cong \mathfrak{M}_3^+$ either as a factor group or as a subgroup. As a factor group, $\mathfrak{P}_3$ is defined by 7.36 with the extra relation

$$(O\,Q)^3 = E.$$

The generators can be represented by the matrices

$$P = \begin{pmatrix} 0 & -1 & 0 \\ -1 & 0 & 0 \\ 0 & 0 & -1 \end{pmatrix}, \; O = \begin{pmatrix} 1 & 0 & 0 \\ 0 & -1 & 0 \\ 0 & 0 & -1 \end{pmatrix}, \; Q = \begin{pmatrix} 0 & 1 & 0 \\ 0 & 0 & 1 \\ 1 & 0 & 0 \end{pmatrix}, \; U = \begin{pmatrix} 1 & 1 & 0 \\ 0 & 1 & 0 \\ 0 & 0 & 1 \end{pmatrix}.$$

Nielsen (1924a, p. 26), regarding $\mathfrak{M}_3^+$ as the subgroup of $\mathfrak{M}_3$ generated by Q, U and $T = PO$, gave the alternative presentation

$$\begin{gathered} Q^3 = T^4 = (Q\,T)^2 = (Q^{-1} T^2\, Q\,U)^2 = E, \; (T^{-1} U)^3 = T^2, \\ T^{-1} U\,T \rightleftarrows Q^{-1} U Q, \; T^{-1} U\,T \rightleftarrows Q\,U\,Q^{-1}, \\ Q\,U\,Q^{-1} U\,Q\,U^{-1}\,Q^{-1} U^{-1} = T^{-1}\,Q\,U\,Q^{-1}\,T. \end{gathered} \tag{7.37}$$

7.4 Linear fractional groups. An important family of finite groups is obtained by using, instead of integers, the marks (elements) of the Galois field $GF(q)$, where $q = p^m$ is a power of a prime. Thus we have the *general linear homogeneous* group $GL(n, q)$, of order

$$\Omega = (q^n - 1)\,(q^n - q)\,(q^n - q^2) \cdots (q^n - q^{n-1}),$$

consisting of $n \times n$ matrices of non-zero determinant (Dickson 1901a, p. 77; van der Waerden 1948, p. 6). In the finite affine n-space $EG(n, q)$, it arises as the group of affine collineations leaving one point invariant. Its centre consists of all matrices μI, where μ is any mark of $GF(q)$ different from zero; i.e., the centre is generated by λI, where λ is a primitive mark of the field. Thus the central quotient group

$$PGL(n, q) \cong GL(n, q)/\{\lambda I\}$$

is of order $\Omega/(q-1)$. The lines of $EG(n, q)$ through a single point constitute a finite projective geometry $PG(n-1, q)$, in which $PGL(n, q)$ is the group of projective collineations. (This is a subgroup of index m in the group of all collineations, since the latter includes the group of automorphisms of $GF(q)$, which is the $\mathfrak{C}_m$ generated by the transformation

$$x \to x^p;$$

see Carmichael 1937, p. 362.) The elements of $PGL(n, q)$ may be expressed as linear fractional transformations in $n-1$ variables.

The matrices of determinant 1 in $GL(n, q)$ form a subgroup of index $q-1$: the *special linear homogeneous* group $SL(n, q)$, of order $\Omega/(q-1)$.

Its centre, consisting of the elements μI where $\mu^n = 1$, is generated by νI where

$$\nu = \lambda^{(q-1)/d}, \quad d = (n, q-1).$$

Since the centre is of order d, the central quotient group

$$PSL(n, q) \cong SL(n, q)/\{\nu I\}$$

is of order $\Omega/d(q-1)$ (DICKSON 1901a, p. 87). This is also the central quotient group of another subgroup of $GL(n, q)$, namely the subgroup of index d consisting of matrices whose determinants are n^{th} powers. (The distinct n^{th} powers in $GF(q)$ are, of course, the first $(q-1)/d$ powers of λ^n.)

To a certain extent, the groups

$$GL(n, q), \quad PGL(n, q), \quad SL(n, q), \quad PSL(n, q)$$

may be regarded as finite counterparts of the infinite groups

$$\mathfrak{M}_n, \quad \mathfrak{P}_n, \quad \mathfrak{M}_n^+, \quad \mathfrak{P}_n^+.$$

In particular, since the Galois field $GF(p)$ is a homomorphic image of the ring of integers, $SL(n, p)$ and $PSL(n, p)$ are factor groups of $\mathfrak{M}_n^+$ and $\mathfrak{P}_n^+$.

The above symbols for the finite groups are due to VAN DER WAERDEN (1948). CARMICHAEL (1937, p. 294) used $P(n-1, q)$ instead of $PGL(n, q)$. For $PSL(n, q)$ we shall follow the majority of authors in using DICKSON's symbol $LF(n, q)$. The following table shows also the notation of DIEUDONNÉ (1955, p. 109):

DICKSON	VAN DER WAERDEN	DIEUDONNÉ
$GLH(n, q)$	$GL(n, q)$	$GL_n(\mathfrak{F}_q)$
	$PGL(n, q)$	$PGL_n(\mathfrak{F}_q)$
$SLH(n, q)$	$SL(n, q)$	$SL_n(\mathfrak{F}_q)$
$LF(n, q)$	$PSL(n, q)$	$PSL_n(\mathfrak{F}_q)$

7.5 The case when $n = 2$ and $q = p$, a prime. As we have agreed to read products from left to right, the linear fractional transformation 7.28, with $ad - bc = 1$, is represented by either of the matrices

$$\begin{pmatrix} a & c \\ b & d \end{pmatrix}, \quad \begin{pmatrix} -a & -c \\ -b & -d \end{pmatrix} \tag{7.51}$$

which, for the moment, we agree to identify. Such transformations (mod p) constitute the linear fractional group $LF(2, p)$, whose order is 6 when $p = 2$, and $p(p^2 - 1)/2$ for any other prime. As generators we naturally choose

$$S: z' = z + 1, \qquad T: z' = -1/z \pmod{p}$$

or

$$S = \begin{pmatrix} 1 & 0 \\ 1 & 1 \end{pmatrix}, \quad T = \begin{pmatrix} 0 & 1 \\ -1 & 0 \end{pmatrix} \tag{7.52}$$

in terms of which

$$\begin{pmatrix} a & c \\ b & d \end{pmatrix} = \begin{cases} S^{(b-1)/a}\, TS^{-a}\, TS^{-(c+1)/a}\, T & (a \neq 0)\,, \\ TS^{-c}\, TS^{b}\, TS^{c(d-1)}\, T & (a = 0)\,. \end{cases}$$

Since these generators satisfy the relations

$$S^p = T^2 = (ST)^3 = E\,,$$

$LF(2, p)$ is a factor group of $(2, 3, p)$ (see 6.41). In the first three cases, the orders 6, 12, 60 show that the homomorphism is an isomorphism:

$$LF(2, p) \cong (2, 3, p) \qquad (p = 2, 3, 5)\,.$$

When $p \geqq 7$, we see from § 5.3 (p. 54) that $(2, 3, p)$ is infinite, so that $LF(2, p)$ needs at least one further relation. The presentations

$$S^7 = T^2 = (ST)^3 = (S^4 T)^4 = E$$

for $LF(2, 7)$ (Dyck 1882, p. 41; Burnside 1911, p. 422) and

$$S^{11} = T^2 = (ST)^3 = (S^4 TS^6 T)^2 = E$$

for $LF(2, 11)$ (Frasch 1933, p. 252; Lewis 1938; see Table 6) suggest the possibility that

$$S^p = T^2 = (ST)^3 = (S^4 TS^{(p+1)/2} T)^2 = E \tag{7.53}$$

may still suffice for greater values of p. (We easily verify that the final relation is satisfied by the generators 7.52.) The truth of this conjecture was established by J. G. Sunday when he succeeded in deducing Mennicke's presentation

$$S^p = T^2 = (ST)^3 = (S^2 TS^{(p+1)/2} T)^3 = E$$

(Behr and Mennicke 1968) from 7.53. Writing k for $(p + 1)/2$, we may sketch his procedure as follows. After proving

$$(TS^2 T)^{-1} S (TS^2 T) = (S^k T) S^4 (S^k T)^{-1}\,,$$

he took kth powers to obtain

$$(TS^2 T)^{-1} S^k TS^2 T = S^k TS^2 (S^k T)^{-1}$$

whence $S^k TS^2 TS^k T = TS^2 TS^k TS^2$ and

$$(S^2 TS^k T)^3 = (S^2 TS^k TS^2)^2 = E\,.$$

ZASSENHAUS (1969) slightly modified MENNICKE's presentation, reducing the number of relations from four to three. However, his presentation

$$S^p = (S\,T)^3\,, \quad T^2 = (S^2\,T\,S^{(p^2+1)/2}\,T)^3 = E$$

fails not only when $p = 3$ or 17 (as he stated) but also for greater values congruent to 3 (mod 14), such as 31 and 59. SUNDAY (1972) obtained

$$S^p = E\,, \quad T^2 = (S\,T)^3\,, \quad (S^4\,T\,S^{(p+1)/2}\,T)^2 = E \qquad (7.54)$$

as a presentation of $LF(2, p)$ for all odd primes p. The crucial step is the observation that these relations, along with $S \rightleftarrows T$, imply $S^3 = T = E$, so that T is a commutator.

A useful presentation for $LF(2, p)$ $(p > 2)$ in terms of S, T and the redundant generator

$$V = \begin{pmatrix} \alpha & 0 \\ 0 & \alpha^{-1} \end{pmatrix}, \qquad \alpha \text{ a primitive root (mod } p),$$

of period $\frac{1}{2}(p - 1)$, is

$$S^p = T^2 = (S\,T)^3 = (T\,V)^2 = (S^\alpha\,T\,V)^3 = E\,, \quad V^{-1}\,S\,V = S^{\alpha^2}\,.$$

Actually, FRASCH (1933, p. 252) used

$$S^p = V^{(p-1)/2} = T^2 = (S\,T)^3 = (T\,V)^2 = E\,, \qquad V^{-1}\,S\,V = S^{\alpha^2} \qquad (7.541)$$

with the extra relation $(S^\alpha\,T\,V)^3 = E$ when $p \equiv 1$ (mod 4) (cf. TODD 1932a, pp. 195—196).

Without the identification 7.51, the matrices 7.52 generate $SL(2, p)$, the group of all 2×2 matrices of integers (mod p) of determinant 1, including

$$Z = \begin{pmatrix} -1 & 0 \\ 0 & -1 \end{pmatrix}.$$

Thus 7.54 leads to the presentation

$$S^p = E\,, \quad T^2 = (S\,T)^3 = (S^4\,T\,S^{(p+1)/2}\,T)^2 \qquad (7.55)$$

for $SL(2, p)$.

Adjoining to $SL(2, p)$ a new element

$$Q = \begin{pmatrix} \alpha & 0 \\ 0 & 1 \end{pmatrix},$$

where α is a primitive root modulo p, we obtain $GL(2, p)$, the group of all 2×2 matrices of integers (mod p) of non-zero determinant. For if A is such a matrix of determinant α^t, then $Q^{-t}A$, having determinant 1, belongs to $SL(2, p)$; thus $GL(2, p)$ is generated by S, T, Q. Since Q, of period $p-1$, transforms S and T into S^α and

$$\begin{pmatrix} 0 & 1/\alpha \\ -\alpha & 0 \end{pmatrix} = T S^{-1/\alpha} T S^{-\alpha} T S^{-1/\alpha} T ,$$

a presentation for $GL(2, p)$ can be derived from **7.55** by adding the three extra relations

$$Q^{p-1} = E , \quad Q^{-1} S Q = S^\alpha , \quad Q^{-1} T Q = T S^{-1/\alpha} T S^{-\alpha} T S^{-1/\alpha} T . \tag{7.56}$$

Since the centre of $GL(2, p)$ is generated by

$$\begin{pmatrix} \alpha & 0 \\ 0 & \alpha \end{pmatrix} = (Q T)^2 Z ,$$

we can derive a presentation for $PGL(2, p)$ by setting $(QT)^2 = Z = E$. Thus $PGL(2, p)$ is given by **7.54** and **7.56** with

$$(Q T)^2 = E . \tag{7.57}$$

Simpler presentations for $PGL(2, p)$ are known in certain special cases. Using $(l, m, n; k)$ to denote the group

$$R^l = S^m = T^n = RST = (TSR)^k = E \tag{7.58}$$

or $R^l = S^m = (RS)^n = (R^{-1} S^{-1} RS)^k = E$, and $G^{p,q,r}$ to denote

$$A^p = B^q = C^r = (AB)^2 = (BC)^2 = (CA)^2 = (ABC)^2 = E \tag{7.59}$$

(Sinkov 1937, p. 68), we have the following abstract definitions (Coxeter 1939, pp. 107—108):

$$PGL(2, 3) \cong [3, 3] \cong \mathfrak{S}_4 \cong G^{3,3,4} ,$$
$$PGL(2, 5) \cong \mathfrak{S}_5 \cong (2, 4, 5; 3) \cong (2, 5, 6; 2) ,$$
$$PGL(2, 7) \cong (2, 3, 8; 4) \cong G^{3,7,8} ,$$
$$PGL(2, 13) \cong (2, 4, 7; 3) \cong G^{3,7,12} \cong G^{3,7,14} ,$$
$$PGL(2, 19) \cong G^{4,5,9} ,$$
$$PGL(2, 23) \cong G^{3,8,11} .$$

It is interesting to observe that, in the same notation,

$$LF(2, 5) \cong (2, 3, 5) \cong G^{3,5,5} ,$$
$$LF(2, 7) \cong (2, 3, 7; 4) ,$$
$$LF(2, 11) \cong G^{5,5,5} ,$$
$$LF(2, 13) \cong (2, 3, 7; 6) \cong (2, 3, 7; 7) \cong G^{3,7,13} ,$$
$$LF(2, 19) \cong (2, 5, 9; 2) \cong G^{3,9,9} ,$$

$$LF(2, 23) \cong (2, 3, 11; 4),$$
$$LF(2, 29) \cong G^{3,7,15}.$$

(COXETER 1939, pp. 89, 94—95, 106, 115[1]), 116; SINKOV 1935, p. 239).

7.6 The groups $LF(2, 2^m)$. BUSSEY (1905, p. 297) gave complicated generating relations for the simple group $LF(2, 2^m)$, $m > 1$, of order $2^m(2^{2m} - 1)$, in terms of two or three generators. TODD (1936, p. 106) discovered a somewhat simpler definition in terms of the $m + 2$ generators

$$R: z' = \alpha z + 1, \quad U: z' = \frac{1}{z+1}, \quad S_i: z' = z + \alpha^i \ (i = 0, 1, \ldots, m-1),$$

where α is a primitive root of $GF(2^m)$. Obviously R, U and $S = S_0$ suffice to generate the group. Using these generators alone, SINKOV (1939) reduced TODD's $\binom{m+2}{2} + 3$ relations to the following $m + 5$:

$$R^{2^m-1} = S^2 = U^3 = (RU)^2 = (SU)^2 = (R^{-i}SR^iS)^2 = E \ (i = 1, 2, \ldots, m-1),$$

$$R^m = SRS^{a_1}R \ldots S^{a_{m-1}}RS^{a_m},$$

where the a's (each 0 or 1) are the coefficients in the irreducible equation

$$\alpha^m + a_1\alpha^{m-1} + \cdots + a_{m-1}\alpha + a_m = 0$$

satisfied by the primitive root α. SINKOV's three generators are not independent: R and U suffice to generate the group (SINKOV 1938, p. 455).

Abstract definitions containing only $m + 1$ relations are known for $m = 2, 3, 4, 5$. In the first case we have simply

$$LF(2, 2^2) \cong (2, 3, 5).$$

BURNSIDE (1899, p. 174) defined $LF(2, 2^3)$ in the form

$$A^7 = B^2 = (AB)^3 = (A^3BA^5BA^3B)^2 = E.$$

SINKOV (1937, p. 70; 1939, pp. 763, 764) found the following definitions:

$$LF(2, 2^3): \ P^7 = (P^2Q)^3 = (P^3Q)^2 = (PQ^5)^2 = E,$$
$$LF(2, 2^4): \ P^{15} = (P^2Q)^3 = (P^3Q)^2 = (PQ^9)^2 = (P^8Q^2)^2 = E,$$
$$LF(2, 2^5): \ P^{31} = (P^2Q)^3 = (P^3Q)^2 = (PQ^6)^2$$
$$= (P^2Q^4P^4Q^5)^2 = (P^2Q^5P^4Q^4)^2 = E.$$

In the notation of 7.59, $LF(2, 2^3) \cong G^{3,7,9}$.

---

[1]) The presentation $G^{3,7,13}$ for $LF(2, 13)$ was established by JOHN LEECH (1962, p. 168). This corrects an error in SINKOV 1937, p. 73, which was repeated in COXETER 1962a, p. 54: in Fig. 2 the points (7, 13) and (13, 7) should be marked 1092 instead of 2184.

7.7 The Hessian group and $LF(3, 3)$. Consider the curve

$$x^3 + y^3 + z^3 + 6mxyz = 0$$

in the complex projective plane. Its points of inflexion, being its intersections with its Hessian

$$x^3 + y^3 + z^3 - \frac{2m^3+1}{m^2} xyz = 0,$$

are given by

$$x^3 + y^3 + z^3 = xyz = 0$$

(WEBER 1896, p. 340). Accordingly they are the nine points

$$\begin{array}{lll} (0, 1, -1), & (0, 1, -\omega), & (0, 1, -\omega^2), \\ (-1, 0, 1), & (-\omega, 0, 1), & (-\omega^2, 0, 1), \\ (1, -1, 0), & (1, -\omega, 0), & (1, -\omega^2, 0), \end{array} \tag{7.71}$$

where $\omega = e^{2\pi i/3}$. When thus arranged as a 3×3 matrix, they are easily seen to lie by threes on twelve lines so as to form a configuration $(9_4, 12_3)$; the twelve lines correspond to the rows, columns, and diagonals (including the "broken" diagonals, as in the terms of the expansion of a determinant). This configuration also represents the finite affine geometry $EG(2, 3)$ of 9 points and 12 lines, each line containing 3 points; hence the group of automorphisms of the configuration is the group of affine collineations of $EG(2, 3)$, of order 432 (CARMICHAEL 1937, pp. 329, 374, 391). The 216 "direct" collineations form the Hessian group of order 216. With the nine points 7.71 numbered in the order

$$\begin{array}{ccc} 1, & 4, & 7 \\ 2, & 5, & 8 \\ 3, & 6, & 9 \end{array}$$

(MILLER, BLICHFELDT and DICKSON, 1916, p. 335), the Hessian group, represented as a doubly transitive permutation group of degree 9, is generated by

$$T = (1\ 2\ 4)\ (5\ 6\ 8)\ (3\ 9\ 7), \quad U = (4\ 5\ 6)\ (7\ 9\ 8).$$

In terms of these, COXETER (1956, p. 168) found the abstract definition

$$T^3 = U^3 = (TU)^4 = E, \quad (TUT)^2 U = U(TUT)^2. \tag{7.72}$$

The subgroup generated by U and $V = TUT^{-1} = (2\ 8\ 5)\ (3\ 6\ 9)$ is transitive on the points $2, 3, \ldots, 9$ and leaves the point 1 fixed, i.e., it is the group $SL(2, 3)$ of order 24 (see § 7.4, p. 92). The relations 7.72 imply

$$U^3 = E, \quad UVU = VUV,$$

(cf. 6.66, 6.69); hence

$$SL(2, 3) \cong \langle 2, 3, 3 \rangle. \tag{7.73}$$

The group of all collineations of $EG(2, 3)$ may be derived from the Hessian group by adjoining an involutory element

$$W = (1\ 2)\ (5\ 6)\ (7\ 9)$$

which transforms both T and U into their inverses. Hence this group, of order 432, is defined by

$$T^3 = U^3 = W^2 = (TU)^4 = (TW)^2 = (UW)^2 = [(TUT)^2UW]^2 = E.$$

In the complex projective plane, it is the group of projective and anti-projective collineations leaving invariant the configuration of the nine inflexions.

The simple group $LF(3, 3)$, of order $5616 = 2^4 \cdot 3^3 \cdot 13$, can be shown to be generated by the matrices

$$S = \begin{pmatrix} 0 & -1 & 0 \\ 1 & 1 & 0 \\ 0 & 0 & 1 \end{pmatrix}, \quad T = \begin{pmatrix} 0 & 1 & 0 \\ 0 & 0 & 1 \\ 1 & 0 & 0 \end{pmatrix} \quad \text{(mod 3)},$$

which satisfy

$$S^6 = T^3 = (ST)^4 = (S^2T)^4 = (S^3T)^3 = E, \quad (TS^2T)^2S^2 = S^2(TS^2T)^2$$

(COXETER 1956, pp. 165, 172). The sufficiency of these relations has been established by enumerating the 26 cosets of the subgroup $\{T, S^2\}$, which is the Hessian group of order 216 in the form 7.72.

7.8 The Mathieu groups. Lists of finite simple groups have been given by DICKSON (1901a; 1905), TITS (1970) and CONWAY (1971, pp. 246–247). With but five exceptions DICKSON's groups fall into infinite families. The five exceptional groups, discovered by MATHIEU (1861; 1873), were further investigated by JORDAN (1870), DE SÉGUIER (1904), ZASSENHAUS (1935) and WITT (1938). In WITT's notation, they are M_{11}, M_{12}, M_{22}, M_{23}, M_{24}. Generators for them may be seen in the book of CARMICHAEL (1937, pp. 151, 263, 288); but only recently have defining relations for all the five Mathieu groups been given.

The smallest Mathieu group M_{11}, of order $7920 = 2^4 \cdot 3^2 \cdot 5 \cdot 11$, is generated by the permutations

$$A = (0\ 1\ 2\ \ldots\ 10), \quad B = (1\ 4\ 5\ 9\ 3)\ (2\ 8\ 10\ 7\ 6), \quad C = (1\ 5\ 4\ 3)\ (2\ 6\ 10\ 7)$$

(cf. CARMICHAEL 1937, p. 151; FRYER 1955, pp. 24, 25). In fact, A and C suffice to generate the group, since $B = C^2A^5C^2A^4C^2A^5$. We proceed to prove that relations

$$\begin{gathered} A^{11} = B^5 = C^4 = (AC)^3 = E, \\ B^{-1}AB = A^4, \quad C^{-1}BC = B^2 \end{gathered} \tag{7.81}$$

provide an abstract definition for M_{11}. Since the above permutations satisfy these relations, it will suffice to show that the order of the abstract group $\mathfrak{G}$ defined by 7.81 is at most 7920. Let $\mathfrak{H}$ be the subgroup of $\mathfrak{G}$ generated by the elements $S = A^3$, $V = B$, $T = C^2$. The relations 7.81 imply

$$C^{-1}B^2C = B^4,\ C^{-1}\cdot C^{-1}BC\cdot C = B^4,\ (C^2B)^2 = E,$$

and

$$\begin{aligned} A^3C^2 &= BAB^{-1}\cdot BC^2B = BAC\cdot CB = BAC\cdot BCB^{-1} \\ &= BC^{-1}A^{-1}C^{-1}A^{-1}\cdot A^{-3}BA\cdot CB^{-1} \\ &= BC^{-1}A^{-1}B^{-2}C^{-1}B\cdot B^{-1}A^{-1}B\,.\,BACB^{-1} \\ &= (B^2ACB^{-1})^{-1}(AC)^{-1}B^2ACB^{-1}, \end{aligned}$$

so that $(A^3C^2)^3 = E$. We thus obtain

$$S^{11} = V^5 = T^2 = (ST)^3 = (TV)^2 = E,\ V^{-1}SV = S^4, \quad (7.82)$$

which are FRASCH's relations 7.541 for $LF(2, 11)$ with $\alpha = 2$. Thus $\mathfrak{H}$ is either $LF(2, 11)$ or a factor group of it. Since $LF(2, 11)$ is simple, this means that $\mathfrak{H}$ is precisely $LF(2, 11)$, of order 660. We now enumerate the cosets of $\mathfrak{H}$ in $\mathfrak{G}$ by the method described in Chapter 2, and find that there are precisely 12 cosets. Hence the order of $\mathfrak{G}$ is $12.660 = 7920$, and our presentation for M_{11} is established.

CONWAY (1971, p. 222) exhibited M_{12} (of order 95040) as a permutation group of degree 12 generated by A,

$$T_1 = B^{-1}C^2B = (1\ 9)\ (2\ 6)\ (4\ 5)\ (7\ 8)$$

and

$$T_2 = (0\ \infty)\ (1\ 10)\ (2\ 5)\ (3\ 7)\ (4\ 8)\ (6\ 9),$$

and established the presentation

$$A^{11} = T_1^2 = T_2^2 = (A\,T_1)^3 = (A\,T_2)^3 = (T_1T_2)^{10} = E\,,$$
$$(T_2T_1)^2A\,(T_1T_2)^2 = A^5\,.$$

These relations easily imply that A, $V = (T_1T_2)^4$ and T_i ($i = 1$ or 2) satisfy 7.82 (with A for S); but the two $LF(2, 11)$ subgroups $\{A, T_i\}$ are thoroughly distinct, since one is maximal while the other is not.

Other presentations of the Mathieu groups have been obtained by MOSER (1958) for M_{11} and M_{12}, MENNICKE and GARBE (1964) for M_{11}, M_{12} and M_{22}, CONWAY (1971) and LEECH (1969) for M_{12}, and TODD (1970) for all of them.

COXETER (1958a; see also TODD 1959) gave a matrix representation of M_{12} of degree 6 over $GF(3)$; it consists of all collineations which leave fixed a certain configuration of 12 points in the finite geometry $PG(5, 3)$. M. HALL (1962) exhibited M_{12} as the automorphism group, modulo a centre of order 2, of a Hadamard matrix of order 12 of a type introduced by PALEY (1933).

Chapter 8

Regular Maps

The branch of topology that deals with regular maps on multiply-connected surfaces may be said to have begun when KEPLER (1619, p. 122) stellated the regular dodecahedron $\{5, 3\}$ to obtain the star-polyhedron $\{^5/_2, 5\}$ (COXETER 1963a, pp. 49, 95—105) which is essentially a map of twelve pentagons on a surface of genus 4. The modern work on this subject received its impetus from two different sources: automorphic functions and the four-colour problem. From the former came two maps on a surface of genus 3: one consisting of 24 heptagons (KLEIN 1879b, pp. 461, 560) and one of 12 octagons (DYCK 1880, pp. 188, 510). From the latter came two maps on a torus: one of seven hexagons (HEAWOOD 1890, p. 334 and Fig. 16) and one of five quadrangles (HEFFTER 1891, p. 491, Fig. 2). The systematic enumeration of regular maps on a surface of genus 2 was begun by ERRÉRA (1922) and completed by THRELFALL (1932a, p. 44). Concerning the torus, it was remarked by BRAHANA (1926, p. 238) that "There is no regular map of 8, 10 or 11 hexagons, no map of 14 hexagons although there are maps of 7, 21 and 28 hexagons". The expression that he sought is our 8.42 (BURNSIDE 1911, p. 418). The first mention of maps on non-orientable surfaces seems to have been by TIETZE (1910).

8.1 Automorphisms. By an *automorphism* of a map (§ 3.2) we mean a permutation of its elements preserving the relations of incidence; for any two corresponding faces or vertices, the edges incident with one correspond to the edges incident with the other in the same (or opposite) cyclic order. The automorphisms clearly form a group, called *the group of the map*. This is the natural generalization of the symmetry group of a polyhedron or tessellation, but metrical ideas are no longer used.

An automorphism is determined by its effect on any one face. For if we fix a face, with all its sides and vertices, then each neighbouring face is likewise fixed; so are the neighbours of each neighbour, and so on.

A map is said to be *regular* if its group contains two particular automorphisms: one, say R, which cyclically permutes the edges that are successive sides of one face, and another, say S, which cyclically permutes the successive edges meeting at one vertex of this face (COX-

ETER 1950, p. 419). Since RS reverses an edge, this agrees with the definition given by BRAHANA (1927, p. 269).

It follows that the group of automorphisms for a regular map is transitive on the N_0 vertices, on the N_1 edges, and on the N_2 faces. The map is said to be of type $\{p, q\}$ if p edges belong to a face and q to a vertex, so that the above automorphisms satisfy 4.42 (and, in general, some further relations independent of these). Clearly

$$qN_0 = 2N_1 = pN_2 \tag{8.11}$$

(THRELFALL 1932a, p. 32), and the dual map is of type $\{q, p\}$.

The "regelmäßige Zellsysteme" of THRELFALL are not all regular in this strict sense. Like ERRÉRA and some other authors, he only requires the same number of sides for every face and the same number of edges at every vertex. For instance, in his Table of 14 maps of genus 2 (THRELFALL 1932a, p. 44), only the second, eighth, ninth, and the last three, are regular in our sense.

The "half-turn" RS reverses an edge by interchanging its two ends and also interchanging the two faces to which it belongs. If there is an automorphism R_1 which interchanges these two vertices without interchanging the two faces, the map is said to be *reflexible* (BALL 1974, p. 130). Then its group is generated by the three "reflections" R_1, $R_2 = R_1R$ and $R_3 = R_2S$, which satisfy 4.32 (and usually other independent relations, too). The fundamental region is obtained by subdividing each face into $2p$ triangles, four of which come together at an edge (see, e.g., KLEIN 1879b, p. 448; FRICKE 1892, p. 458). Thus the order of the group is $4N_1$.

If the surface is non-orientable, any regular map on it has an automorphism which preserves a face while reversing the senses along its sides, and thus reverses the senses of two adjacent faces simultaneously. This automorphism may be identified with R_1. Hence *every regular map on a non-orientable surface is reflexible.* If we try to colour the $4N_1$ triangles alternately white and black, so that R and S preserve the colours, we must eventually reach a conflict, where two adjacent triangles are coloured alike. Taking these to be the two triangles "reflected" into each other by R_1, we deduce that R_1 must somehow be expressible in terms of R and S; i.e.,

$$\{R, S\} \cong \{R_1, R_2, R_3\}$$

(BRAHANA and COBLE 1926, p. 19). We shall consider some particular cases in § 8.6.

When $\chi = 2$, the only regular maps are those listed in 5.11; thus every regular map on a sphere is reflexible. On the other hand, as we shall see in § 8.3, some maps on a torus ($\chi = 0$) are reflexible while others are not. For the latter, R and S generate the whole group, whose order is therefore not $4N_1$, but only $2N_1$. Non-reflexible regular maps on (orientable) surfaces of negative characteristic have been found by J. R. EDMONDS (see COXETER 1963b, p. 466) and F. A. SHERK (1962, pp. 17, 20).

Combining the equations 8.11 and 3.21, we see that a map of type $\{p, q\}$ on a surface of characteristic χ has

$$N_0 = 2pr, \quad N_1 = pqr, \quad N_2 = 2qr, \tag{8.12}$$

where, if $\chi \neq 0$,

$$r = \frac{\chi}{4-(p-2)(q-2)}$$

(BILINSKI 1950, p. 147; 1952, p. 66). If $\chi = 0$, so that $(p-2)(q-2) = 4$, there are infinitely many possible values for r, as we shall see in § 8.3.

8.2 Universal covering. Any map of type $\{p, q\}$, on a surface of characteristic $\chi < 2$, can be "unfolded" so as to form part of the regular tessellation $\{p, q\}$ described in § 5.1. The remaining faces of the tessellation must then be regarded as repetitions of the faces already covered. In other words, the given map is derived from its *universal covering* $\{p, q\}$ by making suitable identifications. (When $\chi = 1$, this is the two-sheeted covering of the projective plane by the sphere: the result of identifyin~ antipodes. But when $\chi \leqq 0$, so that $\{p, q\}$ is infinite, the universal covering has infinitely many sheets.)

The group of the map is thus exhibited as a factor group of either $[p, q]$ or $[p, q]^+$, namely the quotient group of a normal subgroup for which the whole unfolded map serves as a fundamental region. Since the generators A_i of this normal subgroup leave no point invariant, they are translations or glide-reflections (or, when $\chi = 1$, the central inversion). When the surface is orientable, they must be translations, since a glide-reflection reverses sense.

Thus the group of the map has an abstract definition consisting of 4.32 or 4.42 together with the extra relations $A_i = E$ which indicate the necessary identifications (THRELFALL 1932a, pp. 8—14).

8.3 Maps of type {4, 4} on a torus. We have seen that the infinite group **p 4** or $[4, 4]^+$, defined by 4.510, has a normal subgroup of index 4 whose fundamental region is a face of $\{4, 4\}$. This Abelian subgroup **p 1** is generated by the two translations

$$X = STS, \quad Y = S^2 T, \tag{8.31}$$

which satisfy 4.501. Regarding X and Y as unit translations along the Cartesian coordinate axes, we see that $X^x Y^y$ translates the origin to the point (x, y).

The two perpendicular translations $X^b Y^c$ and $X^{-c} Y^b$ generate another subgroup of $[4, 4]^+$, similar (and therefore isomorphic) to that generated by X and Y.

In Fig. 3.5a (p. 25) we obtained a torus (covered with a map having one vertex, two edges, and one face) by identifying opposite sides of a rectangle, which may be taken to be the unit square

$$(1, 0) \quad (0, 0) \quad (0, 1) \quad (1, 1).$$

We can equally well obtain a torus by identifying opposite sides of the larger square

$$(b, c) \quad (0, 0) \quad (-c, b) \quad (b - c, b + c),$$

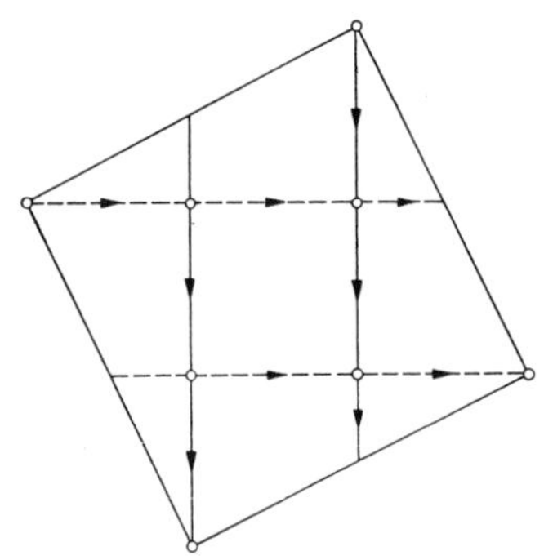

Fig. 8.3a. $\{4, 4\}_{2,1}$ as a Cayley diagram for $\mathfrak{C}_5$

of area $b^2 + c^2$, which is a fundamental region for the group

$$\{X^b Y^c, X^{-c} Y^b\}.$$

On this new torus we find a map

$$\{4, 4\}_{b,c},$$

having n vertices, $2n$ edges, and n faces, where

$$n = b^2 + c^2$$

(Coxeter 1948, p. 25; 1950, p. 421).

Just as the whole $\{4, 4\}$ can be used as a Cayley diagram for the fundamental group $\{X, Y\}$, so $\{4, 4\}_{b,c}$ can be used as a Cayley diagram for the quotient group

$$\{X, Y\}/\{X^b Y^c, X^{-c} Y^b\},$$

which is the group of order n defined by 1.32. (For the case $b = 2$, $c = 1$, see Fig. 8.3a, or Heffter 1891, p. 491. For other Cayley diagrams on a torus, see R. P. Baker 1931.)

The map $\{4, 4\}_{b,c}$ is easily seen to be reflexible if and only if

$$bc(b - c) = 0,$$

in which case its group is

$$[4, 4]/\{X^b Y^c, X^{-c} Y^b\},$$

where, by 5.53 with $p = 1$, $X = A_1 = R_3 R_2 R_1 R_2$ and $Y = A_2 = R_2 R_3 R_2 R_1$. Thus the groups of $\{4, 4\}_{b,0}$ and $\{4, 4\}_{c,c}$ (of orders $8b^2$ and $16c^2$) are given by

$$R_1^2 = R_2^2 = R_3^2 = (R_1 R_2)^4 = (R_2 R_3)^4 = (R_3 R_1)^2 = E \quad (8.32)$$

with the respective extra relations

$$(R_1 R_2 R_3 R_2)^b = E \quad \text{and} \quad (R_3 R_2 R_1)^{2c} = E. \tag{8.321}$$

In particular, $\{4, 4\}_{1,0}$ is the one-faced map that forms the normal decomposition of the torus. By analogy, we may let

$$\{4p, 4p\}_{1,0}$$

denote the one-faced map of genus p considered in § 5.5, p. 58; e.g., $\{8, 8\}_{1,0}$ is shown in Fig. 3.6c. The group of this map, derived from 5.51 and 5.52 by setting $A_1 = E$, is the dihedral group $[4p]$:

$$R_1^2 = R_2^2 = (R_1 R_2)^{4p} = E.$$

On the other hand, the group of the non-reflexible map

$$\{4, 4\}_{b,c} \qquad (b > c > 0)$$

is

$$[4, 4]^+/\{X^b Y^c, X^{-c} Y^b\},$$

of order $4(b^2 + c^2)$, defined by 4.510 with the extra relation (from 8.31)

$$(STS)^b (S^2 T)^c = E$$

(BURNSIDE 1911, p. 416). When $c = 0$ or $b = c$, this group occurs as a subgroup of index 2 in 8.32, 8.321 (DYCK 1880, p. 506).

As we saw in § 4.6, other subgroups of [4, 4] having {4, 4} for their Cayley diagrams are

p 2, p g, p m m, p g g,

defined by 4.502, 4.504, 4.506, 4.508, in terms of which we may take

$$X = T_2 T_3 = P^{-1} Q = R_1 R_3 = O^2, \; Y = T_3 T_1 = P^2 = R_2 R_4 = P^2.$$

In the third and fourth cases, setting $X^b Y^c = X^{-c} Y^b = E$, we obtain two different groups of order $4(b^2 + c^2)$ having $\{4, 4\}_{2b,2c}$ for their Cayley diagrams:

$$R_i^2 = (R_1 R_2)^2 = (R_2 R_3)^2 = (R_3 R_4)^2 = (R_4 R_1)^2$$
$$= (R_1 R_3)^b (R_2 R_4)^c = (R_1 R_3)^{-c} (R_2 R_4)^b = E$$

and

$$(PO)^2 = (P^{-1} O)^2 = O^{2b} P^{2c} = O^{-2c} P^{2b} = E.$$

When $c = 0$, the latter is a special case of the group

$$O^{2a} = P^{2b} = (PO)^2 = (P^{-1} O)^2 = E \tag{8.33}$$

of order $4ab$ (COXETER 1939, p. 81).

In the case of **p 2**, it is natural to make the more general identification $X^b Y^c = X^{b'} Y^{c'} = E \, (bc' - b'c > 0)$ so as to obtain the group

$$T_i^2 = T_1 T_2 T_3 T_4 = (T_2 T_3)^b (T_3 T_1)^c = (T_2 T_3)^{b'} (T_3 T_1)^{c'} = E, \tag{8.34}$$

of order $2(bc' - b'c)$ (BURNSIDE 1911, pp. 410—413). When $b' = -c$ and $c' = b$, the Cayley diagram is $\{4, 4\}_{b+c,b-c}$; in other cases it is an "irregular" map of type $\{4, 4\}$.

In the case of **p g**, where $X = P^{-1}Q$ and $Y = P^2$, we obtain one finite group by setting $X^b = Y^c = E$ and another by merely setting $X^b = Y^c$. The former is the group

$$P^2 = Q^2 = Y, \ (P^{-1}Q)^b = Y^c = E \quad (b > 0, c > 0) \quad (8.35)$$

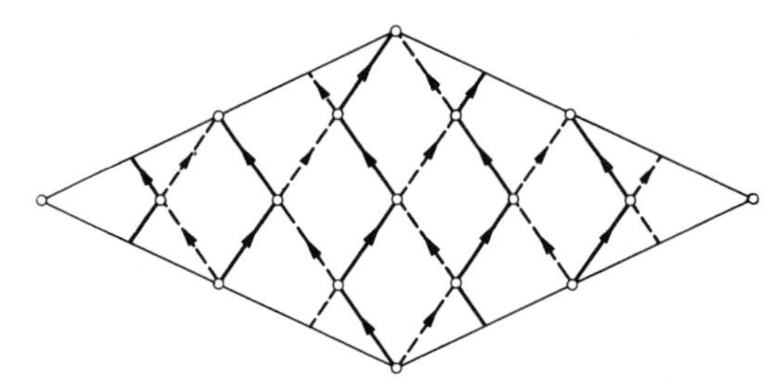

Fig. 8.3b. A Cayley diagram for the dicyclic group $\langle 2, 2, 3 \rangle$ of order 12

of order $2bc$, which is $\langle -2, 2 \mid b; c \rangle$ in the notation of 6.62. Its Cayley diagram is easily drawn by continuing the right half of Fig. 4.5f; when $b = c$, it is $\{4, 4\}_{c,c}$.

Since $PQ^{-1} = P^{-1}Q$, the weaker relations

$$P^2 = Q^2 = Y, \ (P^{-1}Q)^b = Y^c \quad (b > 0, \ c > 0) \quad (8.36)$$

imply $(P^{-1}Q)^{2b} = Y^{2c} = E$, and define a group of order $4bc$. An alternative definition is

$$P^2 = Q^2 = Y, \ (PQ)^b = Y^{b-c}.$$

When $b = c$, the Cayley diagram is $\{4, 4\}_{2b,0}$ (for the case $b = 1$ see the right half of Fig. 4.5f, with opposite sides identified), and the group is simply $\langle 2, 2 \mid b \rangle$, of order $4b^2$, in the form

$$P^2 = Q^2, \ (PQ)^b = E. \quad (8.37)$$

When $b = 1$ the group is $\mathfrak{C}_{4c}$. When $c = 1$, it is the dicyclic group $\langle 2, 2, b \rangle$, of order $4b$, defined by

$$P^2 = Q^2 = (P^{-1}Q)^b \quad (8.38)$$

(cf. 1.64; the Cayley diagram for the case $b = 3$ is shown in Fig. 8.3b). When $c = b - 1$, it is the group $\langle -2, 2, b \rangle$ of order $4b(b - 1)$, defined by

$$P^2 = Q^2 = (PQ)^b, \quad (8.39)$$

which is the direct product $\mathfrak{C}_{b-1} \times \langle 2, 2, b \rangle$ if b is even, but more interesting if b is odd (COXETER 1940c, p. 378).

8.4 Maps of type {3, 6} or {6, 3} on a torus. The infinite group **p 6** or $[3, 6]^+$, defined by 4.516, has a normal subgroup of index 6, whose fundamental region is a face of {6, 3}. This Abelian subgroup **p 1** is generated by the three translations

$$X = S^{-1}TST,\ Y = STSTS,\ Z = TSTS^{-1}, \tag{8.41}$$

which satisfy 4.5011. Regarding X and Y as unit translations along oblique axes inclined at 120°, we see that X^xY^y translates the origin to the point (x, y).

The two translations $X^{b+c}Y^c$ and $X^{-c}Y^b$, likewise inclined at 120° (or the three translations $Y^{-c}Z^b$, $Z^{-c}X^b$, $X^{-c}Y^b$, whose product is E), generate another subgroup of $[3, 6]^+$, similar (and therefore isomorphic) to $\{X, Y\}$ or $\{X, Y, Z\}$.

By identifying opposite sides of the hexagon

$$(1, 0)\ (0, 0)\ (0, 1)\ (1, 2)\ (2, 2)\ (2, 1),$$

we obtain a torus covered with a map having two vertices, three edges, and one face. We can equally well obtain a torus by identifying opposite sides of the larger hexagon

$$(b + c, c)\ (0, 0)\ (-c, b)\ (b - c, 2b + c)\ (2b, 2b + 2c)\ (2b + c, b + 2c),$$

of area $b^2 + bc + c^2$, which is a fundamental region for the group

$$\{X^{b+c}Y^c, X^{-c}Y^b\} \text{ or } \{Y^{-c}Z^b, Z^{-c}X^b, X^{-c}Y^b\}.$$

On this new torus, we find a map

$$\{6, 3\}_{b,c},$$

having $2t$ vertices, $3t$ edges, and t hexagonal faces, where

$$t = b^2 + bc + c^2 \tag{8.42}$$

(Coxeter 1948, p. 25; 1950, p. 421). The dual map

$$\{3, 6\}_{b,c}$$

has (of course) t vertices, $3t$ edges, and $2t$ triangular faces.

Just as the whole {3, 6} can be used as a Cayley diagram for the fundamental group $\{X, Y, Z\}$, so $\{3, 6\}_{b,c}$ can be used as a Cayley diagram for the quotient group

$$\{X, Y, Z\}/\{Y^{-c}Z^b, Z^{-c}X^b, X^{-c}Y^b\},$$

which is the group of order t defined by 1.33 (see Fig. 8.4 for the case $b = 2$, $c = 1$). The map $\{6, 3\}_{2,1}$ was cited by Heawood (1890, p. 334) as one that cannot be coloured with fewer than seven colours (see also Ball 1974, p. 237; Coxeter 1950, p. 424).

The maps $\{3, 6\}_{b,c}$ and $\{6, 3\}_{b,c}$ are reflexible if and only if

$$bc(b - c) = 0,$$

in which case their group is

$$[3, 6]/\{X^{b+c}Y^c, X^{-c}Y^b\},$$

where, if [3, 6] is defined by

$$R_1^2 = R_2^2 = R_3^2 = (R_1R_2)^3 = (R_2R_3)^6 = (R_3R_1)^2 = E, \quad (8.43)$$

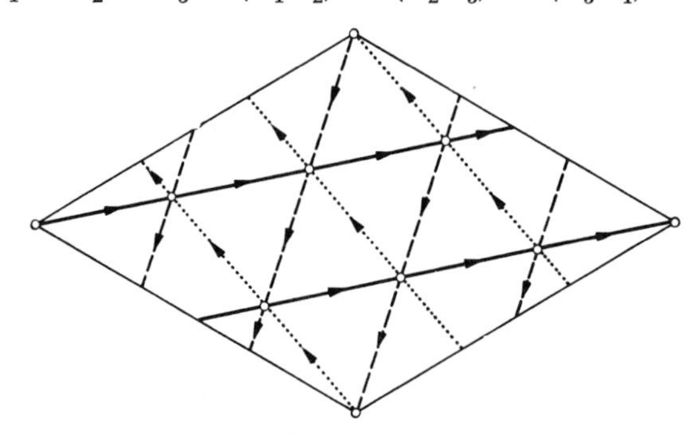

Fig. 8.4. $\{3, 6\}_{2,1}$ as a Cayley diagram for $\mathfrak{C}_7$

we can take $X = (R_3R_2)^2R_1R_2$ and $Y = R_1R_2(R_3R_2)^2$. Thus the groups of $\{3, 6\}_{b,0}$ and $\{3, 6\}_{c,c}$ (of orders $12b^2$ and $36c^2$) are given by 8.43 with the respective extra relations

$$(R_1R_2R_3R_2R_3R_2)^b = E \quad \text{and} \quad (R_1R_2R_3R_2R_3)^{2c} = E. \quad (8.431)$$

On the other hand, the group of the non-reflexible maps

$$\{3, 6\}_{b,c}, \ \{6, 3\}_{b,c} \qquad (b > c > 0)$$

is $[3, 6]^+/\{X^{b+c}Y^c, X^{-c}Y^b\}$, of order $6(b^2 + bc + c^2)$, defined by 4.516 with the extra relation $X^b(XY)^c = E$ or $X^bZ^{-c} = E$, which, by 8.41, is

$$(S^{-1}TST)^b\,(STS^{-1}T)^c = E \quad (8.44)$$

(cf. Burnside 1911, p. 418). When $c = 0$ or $b = c$, this group occurs as a subgroup of index 2 in 8.43, 8.431 (Dyck 1880, p. 506).

As we saw in § 4.6, other subgroups of [3, 6] having {3, 6} or {6, 3} for their Cayley diagrams are

p 3, p 3 m 1,

defined by 4.513, 4.515, in terms of which we may take

$$X = S_1^{-1}S_2 = R_3R_2R_3R_1, \ Y = S_3^{-1}S_1 = R_2R_1R_2R_3,$$

$$Z = S_2^{-1}S_3 = R_1R_3R_1R_2.$$

Setting $X^bZ^{-c} = E$, we obtain the groups

$$S_1^3 = S_2^3 = S_3^3 = S_1S_2S_3 = (S_1^{-1}S_2)^b\,(S_3^{-1}S_2)^c = E, \quad (8.45)$$

of order $3(b^2 + bc + c^2)$ (cf. BURNSIDE 1911, pp. 413—415) and

$$\begin{aligned} R_1^2 = R_2^2 = R_3^2 = (R_1 R_2)^3 = (R_2 R_3)^3 = (R_3 R_1)^3 \\ = (R_3 R_2 R_3 R_1)^b (R_2 R_1 R_3 R_1)^c = E, \end{aligned} \tag{8.46}$$

of order $6(b^2 + bc + c^2)$, having $\{3, 6\}_{b+2c,b-c}$ and $\{6, 3\}_{b+2c,b-c}$ for their respective Cayley diagrams. Note that 8.45 is a subgroup of index 2 in 8.46.

In particular, $\{6, 3\}_{b,b}$ is the Cayley diagram for the group

$$R_1^2 = R_2^2 = R_3^2 = (R_1 R_2)^3 = (R_2 R_3)^3 = (R_3 R_1)^3 = (R_3 R_2 R_3 R_1)^b = E \tag{8.47}$$

of order $6b^2$.

We find similarly that **p 2**, defined by 4.502, yields BURNSIDE's group 8.34 in the form

$$T_1^2 = T_2^2 = T_3^2 = (T_1 T_2 T_3)^2 = (T_2 T_3)^b (T_2 T_1)^c = (T_2 T_3)^{b'} (T_2 T_1)^{c'} = E \tag{8.48}$$

$(bc' - b'c > 0)$, with a Cayley diagram of type $\{6, 3\}$ which is $\{6, 3\}_{b,c}$ when $b' = -c$ and $c' = b + c$.

8.5 Maps having specified holes. In the sense in which the torus is the product of two circles, the regular map $\{4, 4\}_{m,0}$ (consisting of m^2 "squares") is the product of two m-gons. This may be constructed in Euclidean 4-space as a metrically regular *skew polyhedron*, consisting of the m^2 squares faces of the four-dimensional "prism" $\{m\}\times\{m\}$ (COXETER 1937a, p. 43; 1963a, p. 124). Accordingly, we write

$$\{4, 4\}_{m,0} = \{4, 4 \mid m\},$$

and define $\{p, q \mid m\}$ to be an orientable map of type $\{p, q\}$ having m-gonal holes, defined as follows. The *hole* is a path along successive edges, such that at the end of each edge we leave two faces on one side, say the left (the same side at every stage). In other words, $ABCD\ldots$ is a hole if there are faces $A'ABB'\ldots$, $B'BCC'\ldots$, $C'CDD'\ldots$, etc. (COXETER 1937a, p. 37).

The group of $\{p, q \mid m\}$ is easily seen to be 4.32 with

$$(R_1 R_2 R_3 R_2)^m = E$$

(COXETER 1937a, p. 48; cf. 8.321). Its "rotation" subgroup, generated by $R = R_1 R_2$ and $S = R_2 R_3$, is denoted by $(p, q \mid 2, m)$ because it is a special case of the group

$$(p, q \mid r, m)$$

defined by

$$R^p = S^q = (RS)^r = (R^{-1}S)^m = E$$

(COXETER 1939, pp. 74—86, 146).

In particular, $\{3, q \mid q\}$ and $\{q, 3 \mid q\}$ are the ordinary maps $\{3, q\}$ and $\{q, 3\}$, and the infinite maps

$$\{4, 6 \mid 4\},\ \{6, 4 \mid 4\},\ \{6, 6 \mid 3\}$$

are metrically realizable in Euclidean 3-space as regular skew polyhedra (COXETER 1937a, p. 34) which visibly justify our use of the word *hole*. Moreover, $\{5, 5 \mid 3\}$ (BRAHANA and COBLE 1926, p. 14, Fig. 7; THRELFALL 1932a, pp. 19—21) may be identified with POINSOT's star-polyhedron $\{5, {}^5/_2\}$ (COXETER 1963a, p. 95) or with its reciprocal $\{{}^5/_2, 5\}$, discovered by KEPLER (1619, pp. 84—113). The more complicated map $\{5, 4 \mid 4\}$ has been drawn by COXETER (1939, p. 135, Fig. 18).

The only finite maps $\{p, q \mid m\}$ not listed in *Regular skew polyhedra* (COXETER 1937a[1]), p. 61, Table 1) are $\{7, 8 \mid 3\}$, $\{8, 7 \mid 3\}$ (COXETER 1962a, p. 58), and

$$\{p, q \mid 2\},$$

where p and q are both even. Since the group $(p, q \mid 2, 2)$ is of order pq (COXETER 1939, p. 81), $\{p, q \mid 2\}$ is a map of q p-gons having p vertices and $pq/2$ edges. Thus

$$\chi = p + q - pq/2,$$

and the genus is $(i-1)(j-1)$, where $i = p/2$, $j = q/2$.

The definition of a Petrie polygon in § 5.2, extends readily from a regular tessellation to a reflexible map. In particular, the Petrie polygons for dual maps have the same number of sides.

In the case of $\{2i, 2j \mid 2\}$, since the period of R^2S^2 is the least common multiple of i and j, namely $ij/(i, j)$, the Petrie polygon has $2ij/(i, j)$ sides. If i and j are relatively prime, these are just the $2ij$ edges. A cyclic order is thus assigned to the whole set of edges.

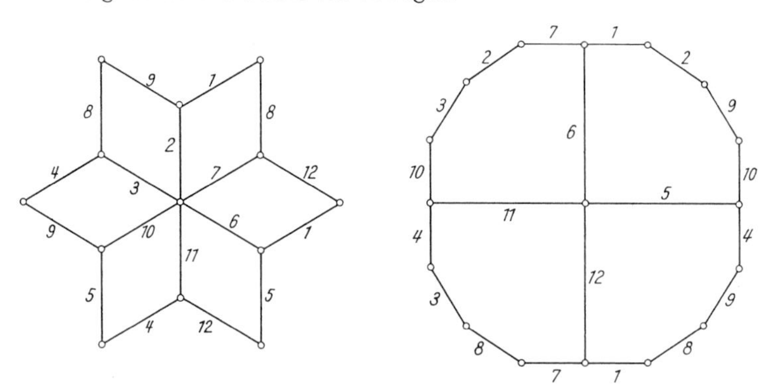

Fig. 8.5. $\{4, 6 \mid 2\}$ and $\{6, 4 \mid 2\}$

[1]) In that paper there is an unfortunate error on p. 55, where the caption for Fig. xv should read $\{4, 6 \mid 3\}$ instead of $\{4, 7 \mid 3\}$.

Fig. 8.5 (cf. ERRÉRA 1922, p. 16, Fig. 24) shows the maps $\{4, 6 \mid 2\}$, $\{6, 4 \mid 2\}$, with their edges numbered consecutively along corresponding Petrie polygons. As permutations of these edges, we have

$$R = (1\ 2\ 7\ 8)\ (3\ 12\ 9\ 6)\ (4\ 5\ 10\ 11),$$

$$S = (1\ 8\ 9\ 4\ 5\ 12)\ (3\ 10\ 11\ 6\ 7\ 2),$$

which are both transformed into their inverses by

$$R_2 = (1\ 8)\ (2\ 7)\ (3\ 6)\ (4\ 5)\ (9\ 12)\ (10\ 11),$$

whence

$$R_3 = R_2 S = (1\ 9)\ (3\ 7)\ (4\ 12)\ (6\ 10),$$

$$R_1 = R R_2 = (1\ 7)\ (3\ 9)\ (5\ 11),$$

and

$$R_1 R_2 R_3 = (1\ 2\ 3\ 4\ 5\ 6\ 7\ 8\ 9\ 10\ 11\ 12).$$

8.6 Maps having specified Petrie polygons. Another fruitful source of regular maps is the identification of those pairs of vertices of $\{p, q\}$ which are separated by r steps along a Petrie polygon. Given p and q, it is an interesting problem to determine suitable values of r. For such a value, the map is denoted by

$$\{p, q\}_r$$

(COXETER 1939, p. 127). Since dual maps have corresponding Petrie polygons, the dual of $\{p, q\}_r$ is $\{q, p\}_r$.

As we saw in § 5.2, the operation $R_1 R_2 R_3$ has the effect of shifting a Petrie polygon one step along itself. Thus we are considering cases where $[p, q]$, defined by 4.32, has a normal subgroup generated by $(R_1 R_2 R_3)^r$ and its conjugates. The map $\{p, q\}_r$ is derived from a fundamental region for this subgroup by making suitable identifications on the boundary, as in Fig. 8.6, p. 113. Its group is the quotient group defined by 4.32 with the extra relation

$$(R_1 R_2 R_3)^r = E. \tag{8.61}$$

Since $R_1 R_2 R_3$ reverses sense, $\{p, q\}_r$ is orientable or non-orientable according as r is even or odd.

When $(p-2)(q-2) < 4$, we have a spherical tessellation $\{p, q\} = \{p, q\}_h$, where h, given by 5.21, is the period of the rotatory-inversion $R_1 R_2 R_3$. When $h/2$ is odd, $(R_1 R_2 R_3)^{h/2}$ is the central inversion Z, and $[p, q]/\{Z\}$ is the group of the map

$$\{p, q\}/2 = \{p, q\}_{h/2},$$

which is derived from $\{p, q\}$ by identifying antipodes. This is a map of regular p-gons filling the elliptic plane, which is topologically the projective plane. We thus obtain the following maps of characteristic 1:

$$\{3, 4\}_3, \ \{4, 3\}_3, \ \{3, 5\}_5, \ \{5, 3\}_5$$

(Hilbert and Cohn-Vossen 1932, Figs. 160—163, 165—168; Coxeter 1950, p. 421). The last of these illustrates the fact the colouring of a map in the projective plane may require as many as six colours (Tietze 1910).

For the sake of completeness, we include also the trivial elliptic maps

$$\{2, 2q\}/2, \ \{2q, 2\}/2$$

(Coxeter 1963b, p. 466, Ex. 2), which are $\{2, 2q\}_q$, $\{2q, 2\}_q$ when q is odd (though not when q is even, in which case $\{2q, 2\}_q = \{q, 2\}$).

Consider a map $\{p, q\}_r$ having N_0 vertices and N_1 edges (see 8.12). Since the various specimens of the Petrie polygon involve every edge twice, these r-gons may be regarded as the faces of a regular map of type $\{r, q\}$ having the same N_0 vertices and the same N_1 edges ,though covering a different surface. The two maps are symmetrically related: the faces of each are the Petrie polygons of the other. Thus the new map is $\{r, q\}_p$. Dualizing, we obtain altogether six related maps:

$$\{p, q\}_r, \ \{r, q\}_p, \ \{q, r\}_p, \ \{p, r\}_q, \ \{r, p\}_q, \ \{q, p\}_r$$

(Coxeter 1939, p. 130). That all six have the same group

$$G^{p,q,r}$$

may be seen by writing

$$A = R_1 R_2, \quad B = R_2 R_3, \quad C = R_3 R_2 R_1,$$
$$R_1 = BC, \quad R_2 = BCA, \quad R_3 = CA,$$

so that the relations 4.32, 8.61 take the symmetrical form 7.59.

In particular, the octahedral group $G^{3,3,4} \cong \mathfrak{S}_4$ is the group of the tetrahedron $\{3, 3\} = \{3, 3\}_4$ (including reflections) and also of either of the elliptic maps $\{3, 4\}_3$, $\{4, 3\}_3$. Again, the icosahedral group $G^{3,5,5} \cong \mathfrak{A}_5$ is the group of the elliptic maps $\{3, 5\}_5$, $\{5, 3\}_5$ and also of the self-dual map $\{5, 5\}_3$ (Fig. 8.6) on a surface of characteristic -3. Just as the two-sheeted covering of $\{3, 5\}_5$ is the icosahedron $\{3, 5\} = \{3, 5 \mid 5\}$, so the two-sheeted covering of $\{5, 5\}_3$ is $\{5, 5 \mid 3\}$, of characteristic -6. In fact, we can obtain a metrical realization of $\{5, 5\}_3$ by identifying antipodal elements of the great dodecahedron $\{5, {}^5/_2\}$, whose Petrie polygon is a skew hexagon (by 5.21 with $p = 5$, $q = {}^5/_2$).

The group $G^{2,2q,2q} \cong \mathfrak{C}_2 \times \mathfrak{D}_{2q}$ of order $8q$ (Coxeter 1939, p. 106) is the group of the dihedron $\{2q, 2\} = \{2q, 2\}_{2q}$ and also of the self-dual map $\{2q, 2q\}_2$ which has 2 vertices, $2q$ edges, 2 faces, and is thus of genus $q-1$. In particular, we have $\{4, 4\}_2 \cong \{4, 4\}_{1,1}$, which generalizes in another direction to yield

$$\{4, 4\}_{2c} = \{4, 4\}_{c,c}.$$

Similarly, $G^{2,2p+1,4p+2} \cong \mathfrak{C}_2 \times \mathfrak{D}_{2p+1}$ is the group of the dihedron $\{2p+1, 2\} = \{2p+1, 2\}_{4p+2}$, of the elliptic one-faced map

$$\{4p+2, 2\}_{2p+1} = \{4p+2, 2\}/2,$$

and of two maps of genus p:

$$\{2p+1, 4p+2\}_2$$

having one vertex and two faces,

$$\{4p+2, 2p+1\}_2$$

having one face and two vertices.

In particular, $\{3, 6\}_2 = \{3, 6\}_{1,0}$, $\{6, 3\}_2 = \{6, 3\}_{1,0}$; these generalize in another direction to yield

$$\{3, 6\}_{2b} = \{3, 6\}_{b,0}, \{6, 3\}_{2b} = \{6, 3\}_{b,0}.$$

(See COXETER 1950, p. 414 for $b = 2$, p. 434 for $b = 3$.)

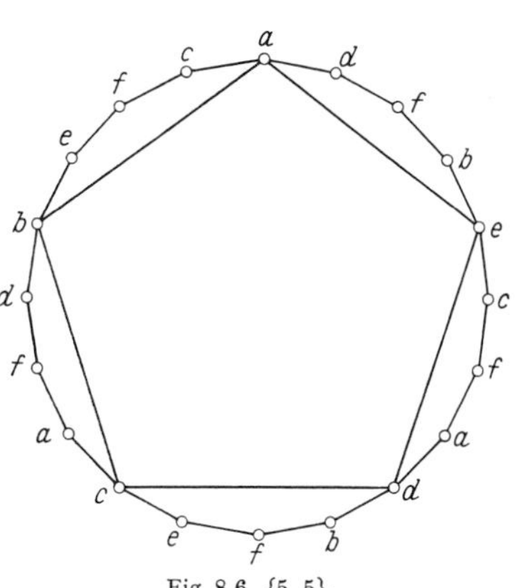

Fig. 8.6. $\{5, 5\}_3$

The complete list of known maps $\{p, q\}_r$ is shown in Table 8 (cf. COXETER 1939, pp. 147—148). The map $\{7, 3\}_8$ was discussed in important papers by KLEIN (1879b, pp. 461, 560) and NIELSEN (1950, pp. 8—12 with $q = 7$). In KLEIN's diagram, two specimens of the Petrie polygon are indicated by means of heavy lines. $\{5, 10\}_3$ is one of the non-orientable dodecahedra of BRAHANA and COBLE (1926, p. 15; Fig. 8); $\{5, 6\}_4$ is the two-sheeted orientable covering for another (ibid., p. 9, Fig. 3). The sixteen pentagons of $\{5, 4\}_5$ have been drawn by COXETER (1939, p. 128, Fig. 8). The only finite groups $G^{p,q,r}$ discovered since 1939 are $G^{3,7,16}$ (COXETER 1962a), $G^{3,7,17}$ (now known to be trivial) and $G^{3,9,10}$ (COXETER 1970, p. 13).

8.7 Maps having two faces. For the following discussion it is desirable to make a slight change of notation so as to agree with BRAHANA (1927): we write

$$S = R_1 R_2, \quad T = R_3 R_1 = R_1 R_3,$$

so that S cyclically permutes the sides of one face while T interchanges this face with one of its neighbours by the "half-turn" that reverses their common side. Thus, for a map of type $\{p, q\}$, S and T satisfy

$$S^p = T^2 = (ST)^q = E.$$

Any regular map yields another (or possibly the same) when we replace its faces by Petrie polygons; e.g., $\{6, 4 \mid 2\}$ (Fig. 8.5) yields the two-faced map of type $\{12, 4\}$ shown in Fig. 8.7a. The "rotation" group of this map is generated by

$$S = (1\ 2\ \ldots\ 12),\ T = (1\ 5)\ (2\ 10)\ (4\ 8)\ (7\ 11).$$

The sides of the second face are cyclically permuted by the transform $TST = S^5$, and the group is

$$S^{12} = T^2 = E,\ TST = S^5,$$

of order 24. By 1.861 with $r = 5$, $m = 3$, $n = 2$, this is $\mathfrak{C}_4 \times \mathfrak{D}_3$.

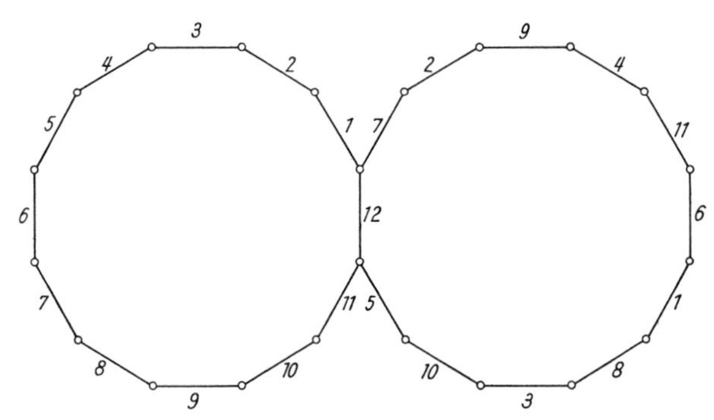

Fig. 8.7a. The map $\{12, 4\}_{1,1}$ of genus 3

More generally, if a regular map of type $\{p, q\}$ has just two faces, its "rotation" group (of order $2p$) is of the form

$$S^p = T^2 = E,\ TST = S^r$$

for a suitable value of r (BRAHANA 1927, p. 280). One face is rotated by S, the other by TST. Since the relations

$$T^2 = E,\ TST = S^r$$

imply

$$S = TS^rT = (TST)^r = S^{r^2},$$

r must satisfy the congruence

$$r^2 \equiv 1 \pmod{p}.$$

When $r = 1$, so that $ST = TS$, we have $\{p, p\}_2$ or $\{p, 2p\}_2$ according as p is even or odd. When $r = -1$ (or $p - 1$), so that $(ST)^2 = E$, we have the dihedron $\{p, 2\}$.

Perhaps the most interesting case is when $p = r^2 - 1$, so that the group, of order $2(r^2 - 1)$, is simply

$$T^2 = E,\ TST = S^{\pm r} \qquad (r > 0) \qquad (8.71)$$

(cf. 1.85). Taking the upper sign, we have $(ST)^2 = S^{r+1}$, so that the period of ST is $2(r-1)$, the group is $\langle -2, r+1 \mid 2 \rangle$ (see § 6.6, p. 74), and the map is of type

$$\{r^2 - 1, 2r - 2\}$$

with $N_0 = r + 1$, $N_1 = r^2 - 1$, $N_2 = 2$, on a surface of genus

$$\frac{(r+1)(r-2)}{2}.$$

Taking instead the lower sign, we have $(ST)^2 = S^{1-r}$, so that the period of ST is $2(r+1)$, the group is $\langle 2, r-1 \mid 2 \rangle$, and the map is of type

$$\{r^2 - 1, 2r + 2\}$$

with $N_0 = r - 1$, $N_1 = r^2 - 1$, $N_2 = 2$, on a surface of genus

$$\frac{r(r-1)}{2}.$$

When $r = 2$, this is $\{3, 6\}_2$; when $r = 3$, it is a map of type $\{8, 8\}$ on a surface of genus 3.

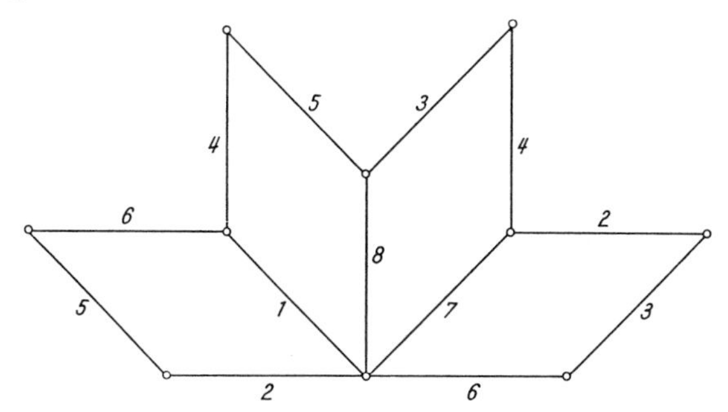

Fig. 8.7b. The map $\{4, 8\}_{1,1}$ of genus 2

The first significant case of 8.71 with the upper sign is

$$T^2 = E, \quad TST = S^3,$$

implying $S^8 = E$; this yields an $\{8, 4\}$ of genus 2. The generalization $S^{4p} = T^2 = E, TST = S^{2p-1}$ or

$$\langle 2, 2p \mid 2; 2 \rangle: \quad S^{2p} = (ST)^2 = Z, \; T^2 = Z^2 = E, \qquad (8.72)$$

of order $8p$, yields a map of type $\{4p, 4\}$ with $N_0 = 2p$, $N_1 = 4p$, $N_2 = 2$, on a surface of genus p (THRELFALL 1932a, p. 46, No. 6). Since this is $\{4, 4\}_{1,1}$ when $p = 1$, an appropriate symbol in the general case is

$$\{4p, 4\}_{1,1}.$$

The dual map $\{4, 4p\}_{1,1}$ (having two vertices) is shown in Fig. 8.7b for the case $p = 2$ (see Table 9 on page 140).

8.8 Maps on a two-sheeted Riemann surface. The group $\langle -3, 4 \mid 2 \rangle$, in BURNSIDE's form 6.672, is the "rotation" group of a map of 6 octagons

(DYCK 1880, p. 488; ERRÉRA 1922, p. 16, Fig. 25; BRAHANA 1927, p. 283), whose 16 vertices and 24 edges can be used as a Cayley diagram for a subgroup of order 16, as in Figs. 3.3c, 3.6c. The dual map, of 16 triangles, has been beautifully drawn by BURNSIDE (1911, pp. 395, 396). It can be derived from the spherical tessellation $\{3, 4\}$ (which is essentially the octahedron) by covering the sphere with a two-sheeted Riemann surface having simple branch-points at the six vertices, namely the Riemann surface for the equation

$$y^2 = x^5 - x$$

(KLEIN 1884, p. 54). Accordingly, natural symbols for these two maps are

$$\{4 + 4, 3\}, \ \{3, 4 + 4\}.$$

They are the two final entries in Table 9, which gives all the regular maps of genus 2.

They generalize to

$$\{p + p, q\}, \ \{q, p + p\},$$

having simple branch-points at the vertices of $\{p, q\}$, which are the face-centres of $\{p, q\}$. Thus the "rotation" group of either map is

$$S^{2p} = T^2 = (S\,T)^q = (S^p\,T)^2 = E \qquad (8.81)$$

(cf. 6.627), where the last relation asserts the commutativity of S^p (which interchanges the two sheets) and T (which reverses an edge).

In the above instance, $p = 4$. Another instance with p even is $\{q, 2 + 2\} = \{q, 4 \mid 2\}$. But when p is odd, 8.81 is the direct product

$$\mathfrak{C}_2 \times [p, q]^+$$

of the group of order 2 generated by S^p and the polyhedral group generated by S^{p+1} and $S^p T$ (COXETER and TODD 1936, p. 196). Thus $\{q, 3 + 3\}$ is $\{q, 6 \mid 2\}$ in the notation of COXETER (1937a, p. 59); e.g.,

$$\{3, 3 + 3\} = \{3, 6\}_4 = \{3, 6\}_{2,0}.$$

Also $\{3, 5 + 5\}$ is the two-sheeted orientable covering of the non-orientable map $\{3, 10\}_5$.

Let us close this section by proving the following theorem:

No regular map can be drawn on a non-orientable surface of characteristic 0 *or* -1.

In the case of characteristic 0 (the Klein bottle), the fundamental group **pg** (Fig. 4.5f) has a special direction determined by the axes of its glide-reflections; therefore it cannot admit a rotation of period greater than 2.

If there were a regular map of characteristic -1, its two-sheeted orientable covering would be a regular map of characteristic -2, that is, of genus 2, and one of the maps in Table 9 would have "opposite" pairs of vertices, and of edges and faces; but this does not happen.

8.9 Symmetrical graphs. A graph (see p. 19) is said to be *symmetrical* if its group of automorphisms is transitive on its vertices and also on its edges. Many such graphs have been obtained by FRUCHT (1955) as Cayley diagrams for groups having involutory generators which are permuted by an outer automorphism. It appears that all the simplest symmetrical graphs can be embedded in suitable surfaces so as to form regular maps. Thus the complete n-point, for $n = 3, 4, 5, 6, 7$, can be embedded in the sphere, the sphere again, the torus, the projective plane, and the torus again, to form

$$\{3, 2\},\ \{3, 3\},\ \{4, 4\}_{2,1},\ \{3, 5\}_5,\ \{3, 6\}_{2,1}.$$

The Thomsen graph (Fig. 3.3d), the Heawood graph (COXETER 1950, p. 424) and the Pappus graph (*ibid.*, p. 434) can be embedded in the torus to form

$$\{6, 3\}_{1,1},\ \{6, 3\}_{2,1},\ \{6, 3\}_{3;0}.$$

The Petersen graph (KÖNIG 1936, p. 194) can be embedded in the projective plane to form $\{5, 3\}_5$. Incidentally, this is the simplest graph which cannot be used as a Cayley diagram.

Finally, as we remarked in § 3.6, p. 30, the Möbius-Kantor graph (Fig. 3.3c) can be embedded in the surface of genus 2 to form $\{4+4, 3\}$ (Fig. 3.6c).

Chapter 9

Groups Generated by Reflections

In this chapter we consider groups whose generators are involutory while their products in pairs have assigned periods; we ask especially what values of the periods will make such a group finite. Cases where the number of generators is less than four have been considered in 4.31 and 4.32. We shall prove that the generators may always be represented by real affine reflections, thus preparing the ground for a complete enumeration of the finite groups. Then we shall describe some of their remarkable properties.

9.1 Reducible and irreducible groups. We consider the general group

$$(R_i R_k)^{p_{ik}} = E \quad (1 \leqq i \leqq k \leqq n,\ p_{ii} = 1) \qquad (9.11)$$

(COXETER 1963a, p. 188), where it is to be understood that $R_i \neq E$, so that, since $p_{ii} = 1$, every generator is of period 2. If we had $p_{ik} = 1$

in any other case, R_k would coincide with R_i; accordingly we shall assume

$$p_{ik} > 1 \qquad (i < k).$$

Such groups were first considered by MOORE (1897) and TODD (1931, p. 224).

When every such $p_{ik} = 2$, so that $R_i R_k = R_k R_i$, we have the direct product of n groups of order 2, that is, the Abelian group of order 2^n and type $(1, \ldots, 1)$. More generally, if the R's fall into two sets such that $p_{ik} = 2$ whenever R_i is in one set and R_k in the other, then the group, being the direct product of groups generated by the separate sets, is said to be *reducible*. On the other hand, if no such reduction occurs, the group is said to be *irreducible*.

9.2 The graphical notation. It is helpful to represent the defining relations 9.11 by a *graph*. (Since this has no connection with the graphs considered in Chapter 3, we shall use the old-fashioned terms *node* and *branch* instead of "vertex" and "edge".) The nodes represent the generators. There is a branch joining the ith and k^{th} nodes whenever $p_{ik} > 2$, the value of p_{ik} being inserted whenever it is greater than 3. When a branch is not so marked, we understand that $p_{ik} = 3$. When two nodes are not (directly) joined, we understand that $p_{ik} = 2$. Thus the graph is connected or disconnected according as the group is irreducible or reducible. In the latter case the group is the direct product of two or more irreducible groups represented by the connected pieces of the graph (COXETER 1935, p. 21).

Some instances of finite irreducible groups have already been mentioned: the p-gonal dihedral group

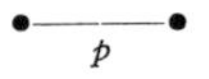

or $[p]$, of order $2p$, and the extended polyhedral groups

[3, 3]	[3, 4]	[3, 5]

of orders 24, 48, 120. Note that [3, 5], though irreducible according to the above definition, is the direct product $\mathfrak{A}_5 \times \mathfrak{C}_2$. Still more surprisingly, as we saw in 1.54, the irreducible group $[2q]$ or $\mathfrak{D}_{2q}$ (q odd) is isomorphic with the reducible group

$$[2, q] \cong \mathfrak{D}_1 \times \mathfrak{D}_q.$$

Thus the question of reducibility depends not only on the abstract group but also on the chosen set of generators.

We have also discussed some infinite groups: the infinite dihedral group

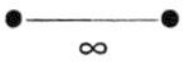

$$\bullet \overset{}{\underset{\infty}{\text{———}}} \bullet \tag{9.21}$$

or $[\infty]$, which is simply $R_1^2 = R_2^2 = E$, and the two-dimensional crystallographic space groups

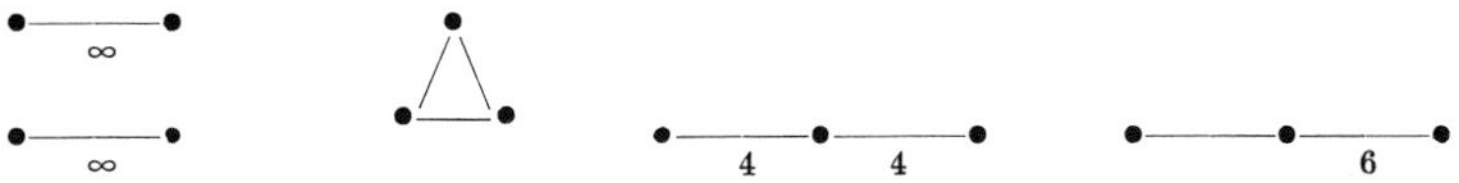

$[\infty] \times [\infty] \cong \mathbf{p\,m\,m}$ $\quad \triangle \cong \mathbf{p\,3\,m\,1}$ $\quad [4, 4] \cong \mathbf{p\,4\,m}$ $\quad [3, 6] \cong \mathbf{p\,6\,m}$

(§ 4.5, pp. 44—51).

9.3 Finite groups. As we saw in § 4.3, the generators of the q-gonal dihedral group

$$R_1^2 = R_2^2 = (R_1 R_2)^q = E$$

may be interpreted as reflections in two lines (of the Euclidean plane) forming an angle π/q. In terms of oblique axes perpendicular to them, the two lines may be expressed as

$$x_1 + a x_2 = 0 \quad \text{and} \quad a x_1 + x_2 = 0$$

where $a = -\cos \pi/q$; and the reflections in them are the affine transformations

$$x_1' = -x_1 - 2 a x_2, \; x_2' = x_2 \tag{9.31}$$

and

$$x_1' = x_1, \; x_2' = -2 a x_1 - x_2. \tag{9.32}$$

These transformations leave invariant the positive definite quadratic form

$$x_1^2 + 2 a x_1 x_2 + x_2^2,$$

which represents the square of the distance from the origin to (x_1, x_2). Their product

$$x_1' = -x_1 - 2 a x_2, \; x_2' = 2 a x_1 + (4a^2 - 1)\, x_2$$

has the characteristic equation

$$(\lambda + 1)^2 - 4 a^2 \lambda = 0,$$

i.e.,

$$\frac{1}{2}(\lambda + \lambda^{-1}) = 2 a^2 - 1 = \cos 2\pi/q;$$

thus its characteristic roots are $e^{\pm 2\pi i/q}$, in agreement with our knowledge that its period is q.

Similarly, let us consider the possibility of representing the generator R_k of the general group 9.11 by the linear transformation

$$x'_k = x_k + 2\left(b_k - \sum_{i=1}^{n} a_{ik} x_i\right),$$

$$x'_j = x_j \qquad (j \neq k),$$

where

$$a_{ik} = a_{ki} = -\cos \pi/p_{ik},$$

so that

$$a_{ii} = 1.$$

When the x's are regarded as coordinates in affine n-space, this is the *affine reflection* in the hyperplane

$$\sum a_{ik} x_i = b_k \qquad (9.33)$$

in the direction of the kth coordinate axis (VEBLEN and YOUNG 1918, pp. 109, 115).

When $\det(a_{ik}) = 0$, the arbitrary constants b_k can be chosen so that the n equations 9.33 have no solution, i.e., so that the n hyperplanes have no common point, but only a common direction. On the other hand, if the group is *finite*, each of its elements leaves invariant the centroid of all transforms of any point in the affine n-space. Hence, for a finite group generated by such affine reflections,

$$\det(a_{ik}) \neq 0 \qquad (9.34)$$

and the invariant point is unique. By translating the origin to this point, we may suppose that

$$b_k = 0,$$

so that R_k is simply

$$x'_k = x_k - 2 \sum a_{ik} x_i, \qquad (9.35)$$

$$x'_j = x_j \qquad (j \neq k).$$

To verify the period of the product of two generators, we note that the product $R_1 R_2$, namely

$$x'_1 = x_1 - 2 \sum a_{i1} x_i,$$

$$x'_2 = x_2 - 2 \sum (a_{i2} - 2 a_{i1} a_{12}) x_i,$$

$$x'_j = x_j \qquad (j > 2),$$

has the characteristic equation

$$\{(\lambda + 1)^2 - 4 a_{12}^2 \lambda\} (\lambda - 1)^{n-2} = 0,$$

whose roots are $\exp(\pm 2\pi i/p_{12})$ and 1 ($n - 2$ times). Thus the affine reflections 9.35 satisfy the relations 9.11, and the geometrical group is isomorphic either with the whole abstract group or with a factor group.

We observe that the transformation 9.35 leaves invariant the definite or indefinite quadratic form

$$\sum \sum a_{ij} x_i x_j, \tag{9.36}$$

which determines a metric in the affine space, two points $(x_1, \ldots, x_n)$ and $(y_1, \ldots, y_n)$ being in perpendicular directions from the origin if

$$\sum \sum a_{ij} x_i y_j = 0. \tag{9.37}$$

(If the form 9.36 is definite, the metric is Euclidean; otherwise it is pseudo-Euclidean, as in MINKOWSKI's space-time; see ROBB 1936.) In the sense of this metric, 9.35 is the ordinary (orthogonal) reflection in the hyperplane

$$\sum_{i=1}^{n} a_{ik} x_i = 0. \tag{9.38}$$

Two such reflections commute if and only if the corresponding hyperplanes are perpendicular.

If all the R_k's leave invariant an m-dimensional subspace through the origin $(1 \leqq m \leqq n-1)$, they also leave invariant the completely orthogonal $(n-m)$-space. Then the reflecting hyperplanes fall into two sets: some containing the m-space and the rest containing the $(n-m)$-space. Because of 9.34, neither set can be empty. Hence our definition of reducibility (§ 9.1, p. 118) agrees with the classical definition of *complete* reducibility for the linear representation in question (BURNSIDE 1911, p. 258).

Any finite group of real linear transformations leaves invariant a positive definite quadratic form: the sum of all the transforms of $\sum x_i^2$. If the group is irreducible, any other invariant quadratic form is merely a numerical multiple of this one, since otherwise we could combine the two to obtain an invariant semidefinite form (the "zero definite" form of BURNSIDE 1911, p. 259), and this would yield an invariant subspace. Hence, if the n R_k's generate a finite irreducible group, the form 9.36 is definite, and the generators are reflections in n concurrent hyperplanes of a *Euclidean n*-space (cf. CARTAN 1928, p. 253).

For a finite reducible group, 9.36 is the sum of several definite forms corresponding to the irreducible components, and again the generators are reflections in n concurrent hyperplanes.

On the unit sphere $\sum \sum a_{ik} x_i x_k = 1$, the n reflecting hyperplanes (perpendicular to the coordinate axes) cut out a spherical simplex whose dihedral angles are π/p_{ik}. Since the spherical $(n-1)$-space is simply-connected, this simplex serves as a fundamental region (COXETER 1963a, pp. 80, 188) and the relations 9.11 suffice to define the geometrical group. Although the Cayley diagram is no longer planar

(§ 3.4, p. 23) when $n > 3$, it still has a set of fundamental circuits which are $2p_{ik}$-gons representing the relations 9.11, as in § 4.3.

On the other hand, whenever the abstract group 9.11 is finite, the corresponding geometrical group is *a fortiori* finite, and the above argument (superseding that of COXETER 1935) establishes their isomorphism:

Every finite group 9.11 *is generated by reflections in the bounding hyperplanes of a spherical simplex.*

Such groups were completely enumerated by COXETER (1931, 1934a, 1963a) and WITT (1941). Those which satisfy the crystallographic restriction ($p_{ik} = 2, 3, 4$ or 6; see COXETER 1958b, p. 319) were enumerated earlier by CARTAN (1927, pp. 218—224). The conclusion is that the irreducible groups are those listed in Table 10 on page 141, and that all others are direct products of these (allowing repetitions), the corresponding graphs being juxtaposed without connection. We observe that the graph for an irreducible finite group is always a tree (connected, but free from circuits, so that the number of branches is one less than the number of nodes, as we remarked on p. 19).

In each case, the order is easily computed by the method of Chapter 2. We merely have to enumerate:

the 24 cosets of $[3, 4]$ in $[3, 4, 3]$,

the 120 cosets of $[3, 5]$ in $[3, 3, 5]$,

the 27 cosets of $[3^{2,1,1}]$ in $[3^{2,2,1}]$,

the 56 cosets of $[3^{2,2,1}]$ in $[3^{3,2,1}]$,

the 240 cosets of $[3^{3,2,1}]$ in $[3^{4,2,1}]$.

The meaning of these symbols is fairly clear from Table 10; e.g., $[3^3]$ is an abbreviation for $[3, 3, 3]$, and $[3^2, 4]$ for $[3, 3, 4]$. To remove any possible doubt, $[p, q, r]$ is

$$\begin{aligned} R_i^2 &= (R_1 R_2)^p = (R_2 R_3)^q = (R_3 R_4)^r \\ &= (R_1 R_3)^2 = (R_1 R_4)^2 = (R_2 R_4)^2 = E \end{aligned} \tag{9.39}$$

or

$$R_i^2 = (R_1 R_2)^p = (R_2 R_3)^q = (R_3 R_4)^r = E,$$

$$R_1 \rightleftarrows R_3,\ R_4; \qquad R_2 \rightleftarrows R_4,$$

and $[3^{2,2,1}]$ is

$$R_i^2 = (R_1 R_2)^3 = (R_2 R_3)^3 = (R_3 R_4)^3 = (R_4 R_5)^3 = (R_3 R_6)^3 = E,$$

$$R_1 \rightleftarrows R_3,\ R_4,\ R_5,\ R_6;\ R_2 \rightleftarrows R_4,\ R_5,\ R_6;\ R_3 \rightleftarrows R_5;\ R_4,\ R_5 \rightleftarrows R_6.$$

9.4 A brief description of the individual groups. As we saw in 6.22, $[3^{n-1}]$ is the symmetric group $\mathfrak{S}_{n+1}$, generated by the transpositions

$$(1\ 2),\ (2\ 3),\ \ldots,\ (n\ n+1).$$

Geometrically, it is the symmetry group of the regular simplex α_n (COXETER 1963a, pp. 121, 133). When $n = 1$, we have the group of order 2, the graph consists of a single node, and the abbreviated symbol is simply [] or [1].

$[3^{n-2}, 4]$, the hyper-octahedral group of § 7.3, p. 90, is the symmetry group of the hyper-cube γ_n, or of the reciprocal cross-polytope β_n, or of the n-dimensional Cartesian frame formed by n mutually perpendicular lines through a point. Its generators are the consecutive transpositions of these n axes and the reversal of the last one. In the notation of PÓLYA (1937, p. 178) it is $\mathfrak{S}_n[\mathfrak{C}_2]$: "der $\mathfrak{C}_2$-Kranz um $\mathfrak{S}_n$".

$[3^{n-3,1,1}]$, a subgroup of index 2 in the hyper-octahedral group, is the symmetry group of the half-measure-polytope $\mathrm{h}\gamma_n$ (COXETER 1963a, p. 155) whose vertices are alternate vertices of γ_n (e.g., $\mathrm{h}\gamma_3$ is the regular tetrahedron). It is also the group of automorphisms of the configuration $(2^{n-1})_n$ of HOMERSHAM COX (1891, p. 67) which is derived from n planes of general position through a point in ordinary space, with one point arbitrarily chosen on each of their $n(n-1)/2$ lines of intersection. There are altogether 2^{n-1} points and 2^{n-1} planes, with n of the points on each plane, and n of the planes through each point. When reciprocities as well as automorphisms are considered, the group is $[3^{n-2}, 4]$ (COXETER 1950, p. 447).

As we saw in 4.31, $[q]$ is the dihedral group $\mathfrak{D}_q$, of order $2q$.

$[3, q]$ $(q = 3, 4, 5)$ are the extended polyhedral groups of KLEIN (1884, p. 23). In terms of the rotatory inversion $R = R_1R_2R_3$ and the rotation $S = R_3R_2$ (COXETER and TODD 1936, p. 195; cf. 7.29) they are defined by

$$S^q = (RS)^2 = (R^2S^2)^3 = (R^3S^2)^2 = E.$$

(One easily verifies that $R_3 = R^2S^2R$.) When $q = 4$ or 5, there is a simpler definition (8.81 with $p = 3$) making use of the expression of the group as a direct product. When $q = 3$, the group is $\mathfrak{S}_4$, and the above relations are easily seen to be equivalent to

$$R^4 = S^3 = (RS)^2 = E.$$

$[3, 4, 3]$ is the symmetry group of the self-reciprocal 24-cell $\{3, 4, 3\}$ in Euclidean 4-space (COXETER 1963a, p. 149). In terms of $P = R_1R_3R_4$ and $Q = R_1R_2R_4$, this is

$$P^6 = Q^6 = (PQ)^4 = (P^3Q)^2 = (PQ^3)^2 = E.$$

[3, 3, 5] is the symmetry group of two remaining regular four-dimensional polytopes: the 600-cell {3, 3, 5} and its reciprocal, the 120-cell {5, 3, 3} (*ibid.*, p. 153). In terms of P and Q,

$$P^{10} = Q^6 = (PQ)^3 = (P^5Q)^2 = (PQ^3)^2 = E.$$

$[3^{2,2,1}]$ is the group of automorphisms of the 27 lines on the general cubic surface in projective 3-space (SCHLÄFLI 1858; TODD 1932b; COXETER 1940b).

In the case of $[3^{n-4,2,1}]$ $(5 \leqq n \leqq 8)$ the invariant quadratic form 9.36 can be expressed as

$$x_1^2 - x_1x_2 + x_2^2 - x_2x_3 + \cdots - x_{n-2}x_{n-1} + x_{n-1}^2 - x_{n-3}\,x_n + x_n^2 \quad (9.41)$$

(COXETER 1951a, p. 405). In terms of new variables

$$y_1 = x_n,\ y_2 = x_{n-1} + 2x_n,\ y_3 = x_{n-2} + x_n,\ y_4 = x_{n-3}, \ldots, y_n = x_1,$$

we have the equivalent form

$$4y_1^2 - 3y_1y_2 + y_2^2 - y_2y_3 + y_3^2 - \cdots + y_{n-1}^2 - y_{n-1}y_n + y_n^2, \quad (9.42)$$

which serves to identify $[3^{n-4,2,1}]$ with the group of linear transformations considered by BURNSIDE (1912, p. 300). The same forms 9.42 ($n = 5$, 6, 7, 8) were obtained by O'CONNOR and PALL (1944, p. 329) as integral positive forms of minimum determinant. They are equivalent (COXETER 1951a, pp. 394, 418, 434) to the extreme forms

$$V_5, X, Y, W_8$$

of KORKINE and ZOLOTAREFF (1873), which were shown by BLICHFELDT (1929) to be "absolutely" extreme.

For greater values of n, the group $[3^{n-4,2,1}]$ is infinite, and the form is semidefinite ($n = 9$) or indefinite ($n > 9$). The groups $[3^{4,2,1}]$ and $[3^{5,2,1}]$ have an interesting application to the algebra of Cayley numbers or *octaves* (COXETER 1946b, p. 571).

9.5 Commutator subgroups. If the words in the defining relations for a given abstract group are such that each involves an even number of letters, then every element is either "even" or "odd" according to the parity of the number of letters occurring in any expression for it. The "even" elements form a subgroup of index 2.

In the case of 9.11, one possible set of $n - 1$ generators for this subgroup is

$$S_i = R_n R_i \qquad (1 \leqq i < n),$$

in terms of which

$$S_i^{p_{in}} = E \qquad (1 \leqq i < n),$$

$$(S_i^{-1} S_k)^{p_{ik}} = E \qquad (1 \leqq i < k < n).$$

When the group is irreducible, another possible set of generators for the subgroup corresponds in a natural manner to the set of branches of the graph: a branch joining the ith and k^{th} nodes yields the generator $R_i R_k$. If its period p_{ik} is odd, this is a commutator; e.g., if $p_{ik} = 3$,

$$R_i R_k = R_k R_i R_k R_i,$$

and if $p_{ik} = 5$,

$$R_i R_k = R_k (R_i R_k R_i) \, R_k (R_i R_k R_i) .$$

Since any commutator

$$(R_l \ldots R_k) \, (R_j \ldots R_i) \, (R_k \ldots R_l) \, (R_i \ldots R_j)$$

involves an even number of R's, we easily deduce:

9.51 *When every branch of the graph is either unmarked or marked with an odd number, the "even" subgroup is the commutator subgroup.*

Following Dieudonné (1955, p. 48), we use the symbol $\mathfrak{G}^+$ for the "even" subgroup of $\mathfrak{G}$. Thus the elements $T_i = R_i R_{i+1}$ of 9.39 generate $[p, q, r]^+$ (Todd 1931, p. 217) in the form

$$T_1^p = T_2^q = T_3^r = (T_1 T_2)^2 = (T_1 T_2 T_3)^2 = (T_2 T_3)^2 = E .$$

It follows from 9.51 that the commutator subgroups of

$$[3^{n-1}], \ [3^{n-3,1,1}], \ [2p+1], \ [3, 5], \ [3, 3, 5], \ [3^{n-4,2,1}] \quad (n = 6, 7, 8)$$

are respectively

$$[3^{n-1}]^+, \ [3^{n-3,1,1}]^+, \ [2p+1]^+, \ [3, 5]^+, \ [3, 3, 5]^+, \ [3^{n-4,2,1}]^+ .$$

Evidently $[3^{n-1}]^+$ is the alternating group $\mathfrak{A}_{n+1}$, $[q]^+$ is the cyclic group $\mathfrak{C}_q$, $[3, 5]^+$ is the icosahedral group $\mathfrak{A}_5$, and $[3, 3, 5]^+$ is the rotation group of the regular 600-cell $\{3, 3, 5\}$. In terms of $R = R_1 R_2 R_3 R_4$ and $S = R_2 R_3 R_4 R_1$, $[3, 3, 5]^+$ has the defining relations

$$R^{15} = S^{15} = (R^3 S^5)^2 = (R^5 S^3)^2 , \quad R^3 \rightleftarrows S^3 , \quad R^5 \rightleftarrows S^5$$

(Coxeter 1937b, pp. 316, 322).

Dickson (1901a, pp. 293—299) considered the groups $[3^{2,1,1}]^+$ and $[3^{2,2,1}]^+$, showing that the latter is the simple group of order 25920:

$$[3^{2,2,1}]^+ \cong HA(4, 2^2) \cong FO(5, 3) \cong A(4, 3)$$

(H. F. Baker 1946) or, in the notation of Dieudonné (1954, pp. 312, 313),

$$PU_4^+(\mathfrak{F}_4) \cong PSp_4(\mathfrak{F}_3) .$$

A concise abstract definition (Coxeter 1959a, p. 105) is

$$R^5 = S^4 = (R^{-1}S)^3 = E, \ R^2 S R^2 S R^2 = SRS .$$

Theorem 9.51 covers all the irreducible finite groups 9.11 except $[3^{n-2}, 4]$, $[q]$ (q even), and $[3, 4, 3]$. We proceed to prove that, in each of these remaining cases, the commutator subgroup is of index 4.

Since the commutator subgroup is generated by the commutators of the generators and their conjugates, it is a normal subgroup whose quotient group (the commutator quotient group) is derived from the original group by making the generators commute. Thus the commutator quotient group of 9.11 is

$$(R_i R_k)^{p_{ik}} = E, \quad R_i R_k = R_k R_i,$$

which means that $R_i^2 = E$, while $R_i = R_k$ whenever p_{ik} is odd; e.g., for the groups that we have just been considering, it is $R_1^2 = E$ (with $R_1 = R_2 = \cdots = R_n$), of order 2. Hence the index of the commutator subgroup, being the order of the commutator quotient group, is 2^c, where c is the number of pieces into which the graph falls when any branches that have even marks are removed. When the graph is a tree, c is one more than the number of such even branches. Thus, for $[2p]$ or $[3^{n-2}, 4]$ or $[3, 4, 3]$, we have $c = 2$, $2^c = 4$.

The commutator subgroup of the dihedral group $[2p]$ (of order $4p$) is evidently the cyclic group $\mathfrak{C}_p$.

In the case of $[3^{n-2}, 4]$, defined by

$$R_i^2 = (R_1 R_2)^3 = (R_2 R_3)^3 = \cdots = (R_{n-2} R_{n-1})^3 = (R_{n-1} R_n)^4 = (R_j R_k)^2 = E \quad (j + 1 < k),$$

we observe that the subgroup $[3^{n-3,1,1}]$ is generated by $R_1, \ldots, R_{n-1}$ and $R_n R_{n-1} R_n$, while $[3^{n-3,1,1}]^+$ is generated by the commutators

$$R_1 R_2, R_2 R_3, \ldots, R_{n-2} R_{n-1} \quad \text{and} \quad R_{n-1} R_n R_{n-1} R_n.$$

Since every commutator in $[3^{n-2}, 4]$ involves R_n an even number of times, it follows that $[3^{n-3,1,1}]^+$ is the commutator subgroup of $[3^{n-2}, 4]$. (Thus $[3^{n-2}, 4]$ and $[3^{n-3,1,1}]$ both have the same commutator subgroup.)

In the group $[3, 4, 3]$, defined by

$$R_i^2 = E, (R_1 R_2)^3 = (R_2 R_3)^4 = (R_3 R_4)^3 = (R_1 R_3)^2 = (R_1 R_4)^2 = (R_2 R_4)^2 = E,$$

the elements $S = R_1 R_2$ and $T = R_3 R_4$, which satisfy

$$S^3 = T^3 = (S^{-1} T^{-1} S T)^2 = E, \tag{9.52}$$

generate a subgroup of index 4, say $[3^+, 4, 3^+]$ (Coxeter 1936a, pp. 68, 70). Since S and T are commutators, this is the commutator subgroup of $[3, 4, 3]$. It can be proved (Sinkov 1936, pp. 76—78) that the relations 9.52 suffice for an abstract definition.

More generally (Coxeter 1936a, p. 69), the group $[l^+, 2p, m^+]$, defined by

$$S^l = T^m = (S^{-1} T^{-1} S T)^p = E, \tag{9.53}$$

is a subgroup of index 4 in $[l, 2p, m]$. In particular, as we saw in 4.44,

$$[l^+, 2p, 2^+] \cong [l^+, 2p].$$

9.6 Central quotient groups. It is known (COXETER 1934a, pp. 606 to 618) that each of the groups

$$[3^{n-2}, 4],\ [3^{n-3,1,1}]\ (n \text{ even}),\ [2p],\ [3, 5],$$

$$[3, 4, 3],\ [3, 3, 5],\ [3^{3,2,1}],\ [3^{4,2,1}]$$

contains the *central inversion*, which reverses every vector in the Euclidean n-space. Moreover, this element (which commutes with every other element) can be expressed as

$$Z = (R_1 R_2 \ldots R_n)^{h/2}, \tag{9.61}$$

where the respective values of $h/2$ are

$$n, \quad n-1, \quad p, \quad 5,$$

$$6, \quad 15, \quad 9, \quad 15.$$

Z generates a normal subgroup of order 2 (the *centre*) whose quotient group is defined by 9.11 together with

$$(R_1 R_2 \ldots R_n)^{h/2} = E.$$

When the group $\mathfrak{G}$ is regarded as operating in a spherical $(n-1)$-space, the central quotient group

$$\mathfrak{G}/\mathfrak{C}_2,$$

where $\mathfrak{C}_2$ is $\{Z\}$, operates analogously in the elliptic $(n-1)$-space which is derived by identifying antipodes (COXETER 1934a, p. 616). We can regard the x's of 9.35 as homogeneous coordinates in a projective $(n-1)$-space with an elliptic metric introduced by means of the condition 9.37 for points (x) and (y) to be "perpendicular" (COXETER 1957a, p. 110). Then 9.35 is the harmonic homology whose centre is the kth vertex of the simplex of reference while its axial hyperplane is the polar hyperplane of that vertex. In other words, $\mathfrak{G}/\mathfrak{C}_2$ is the collineation group corresponding to the linear group (or group of matrices) $\mathfrak{G}$.

When $nh/2$ is odd, so that Z is the product of an odd number of R's, the coset of $\mathfrak{G}^+$ in $\mathfrak{G}$ is $\mathfrak{G}^+ Z$. Hence in this case

$$\mathfrak{G} \cong \mathfrak{G}^+ \times \mathfrak{C}_2$$

(where $\mathfrak{C}_2$ is the group of order 2 generated by Z) and

$$\mathfrak{G}/\mathfrak{C}_2 \cong \mathfrak{G}^+.$$

In particular, the icosahedral group is $[3, 5]/\mathfrak{C}_2 \cong [3, 5]^+$. Another instance is

$$[3^{3,2,1}]/\mathfrak{C}_2 \cong [3^{3,2,1}]^+,$$

the group of automorphisms of the 28 bitangents of the general quartic curve in the projective plane (COXETER 1928, pp. 7—9). From the theory of theta-functions (DU VAL 1933, p. 58) this is seen to be the simple group of order $4 \cdot 9! = 1451520$, i.e., the $A(6, 2)$ of DICKSON (1901a, pp. 89, 100, 308).

When n is odd, both $[3^{n-2}, 4]^+$ and $[3^{n-3,1,1}]$ lack the central inversion. Hence $[3^{n-2}, 4]$ is the direct product of either subgroup with $\{Z\}$, and the two subgroups are isomorphic:

$$[3^{n-2}, 4]/\mathfrak{C}_2 \cong [3^{n-2}, 4]^+ \cong [3^{n-3,1,1}] \qquad (n \text{ odd})$$

(COXETER 1936b, p. 295). When $n = 3$, these are different ways of presenting the octahedral group

$$[3, 4]/\mathfrak{C}_2 \cong [3, 4]^+ \cong [3, 3] \cong \mathfrak{S}_4.$$

On the other hand, when $\mathfrak{G}$ is one of the groups

$$[2p], [3^{n-3,1,1}] \ (n \text{ even}), [3^{n-2}, 4] \ (n \text{ even}),$$

$$[3, 4, 3], [3, 3, 5], [3^{4,2,1}],$$

the element Z, involving an even number of R's, belongs to $\mathfrak{G}^+$; thus we can define a new group

$$\mathfrak{G}^+/\mathfrak{C}_2,$$

either as the central quotient group of $\mathfrak{G}^+$ or as the "even" subgroup of $\mathfrak{G}/\mathfrak{C}_2$. Its defining relations are derived from those of $\mathfrak{G}^+$ by adding

$$Z = E.$$

For instance, $[2p]/\mathfrak{C}_2 \cong [p]$, $[2p]^+ \cong \mathfrak{C}_{2p}$, $[2p]^+/\mathfrak{C}_2 \cong [p]^+ \cong \mathfrak{C}_p$.

Similarly, the central quotient group of $[3^+, 4, 3^+]$, or the commutator subgroup of $[3, 4, 3]/\mathfrak{C}_2$, is

$$[3^+, 4, 3^+]/\mathfrak{C}_2,$$

the direct product of two tetrahedral groups (COXETER 1936a, p. 72). Another case of splitting is

$$[3, 3, 5]^+/\mathfrak{C}_2,$$

the direct product of two icosahedral groups.

GOURSAT (1889, pp. 66—79) enumerated the groups of congruent transformations in elliptic 3-space[1]), numbering them from I to LI.

---

[1]) But he accidentally omitted one of the possible ways of combining two dihedral groups. This was pointed out by THRELFALL and SEIFERT (1931, p. 14) and by HURLEY (1951, p. 652). HURLEY found 222 crystallographic point groups in Euclidean 4-space (analogous to the 32 in 3-space which we mentioned in § 4.2). Some progress towards an extension to n dimensions was made by HERMANN (1949).

We easily recognize eleven of them:

$$\mathrm{XX} \cong [3^+, 4, 3^+]/\mathfrak{C}_2,$$

$$\mathrm{XXII} \cong [3^{1,1,1}]^+/\mathfrak{C}_2, \qquad \mathrm{XLII} \cong [3^{1,1,1}]/\mathfrak{C}_2,$$

$$\mathrm{XXVII} \cong [3, 3, 4]^+/\mathfrak{C}_2, \qquad \mathrm{XLVII} \cong [3, 3, 4]/\mathfrak{C}_2,$$

$$\mathrm{XXVIII} \cong [3, 4, 3]^+/\mathfrak{C}_2, \qquad \mathrm{XLV} \cong [3, 4, 3]/\mathfrak{C}_2,$$

$$\mathrm{XXX} \cong [3, 3, 5]^+/\mathfrak{C}_2, \qquad \mathrm{L} \cong [3, 3, 5]/\mathfrak{C}_2,$$

$$\mathrm{XXXII} \cong [3, 3, 3]^+ \cong \mathfrak{A}_5, \qquad \mathrm{LI} \cong [3, 3, 3] \cong \mathfrak{S}_5.$$

It is remarkable that XXVII and XLII are isomorphic (COXETER 1936b, p. 296).

The group $[3, 4, 3]/\mathfrak{C}_2$ (GOURSAT's XLV) is of special interest as being the collineation group that leaves invariant REYE's configuration 12_6, whose 12 points form a desmic system of three tetrahedra (STEPHANOS 1879; HUDSON 1905, pp. 1—3; COXETER 1950, p. 453; 1954, pp. 478, 480), while its 12 planes form another ("conjugate") desmic system.

DU VAL (1933, p. 49) has shown that $[3^{4,2,1}]/\mathfrak{C}_2$, of order $96 \cdot 10!$, is the group of automorphisms of the 120 tritangent planes of the sextic curve of intersection of a cubic surface and a quadric cone. It follows that $[3^{4,2,1}]^+/\mathfrak{C}_2$ is the simple group

$$FH(8, 2) \cong \Omega_8(\mathfrak{F}_2, Q),$$

of order $48 \cdot 10!$ (DICKSON 1901a, pp. 201, 216; DIEUDONNÉ 1955, p. 64).

The groups $[3^{2,2,1}]$, $[3^{3,2,1}]/\mathfrak{C}_2$ and $[3^{4,2,1}]/\mathfrak{C}_2$ were considered by MITCHELL (1914) as collineation groups in real projective spaces of 5, 6 and 7 dimensions (see also HAMILL 1951; EDGE 1963).

9.7 Exponents and invariants. For each of the groups in Table 10, the properties of the element

$$R_1 R_2 \ldots R_n,$$

of period h, are unique, in the sense that by naming the R's in a different order we merely obtain a conjugate element (COXETER 1934a, p. 602). Its characteristic roots are powers of $\varepsilon = e^{2\pi i/h}$, say

$$\varepsilon^{m_j} \qquad (j = 1, \ldots, n;\ m_1 \leqq m_2 \leqq \cdots \leqq m_n = h - 1)$$

(KILLING 1888, p. 20; COXETER 1951b, pp. 768—771). To express the exponents m_j in terms of the numbers $a_{ik} = -\cos \pi/p_{ik}$, we have the equation

$$\begin{vmatrix} X & a_{12} & a_{13} & \cdots & a_{1n} \\ a_{21} & X & a_{23} & \cdots & a_{2n} \\ & \cdots & \cdots & & \\ a_{n1} & a_{n2} & a_{n3} & \cdots & X \end{vmatrix} = 0$$

whose roots are

$$\cos m_j \pi/h \qquad (j = 1, 2, \ldots, n).$$

The Z of 9.61, having characteristic roots

$$e^{m_j \pi i} = (-1)^{m_j},$$

is the central inversion whenever the m_j are all odd. Conversely, the groups that lack the central inversion are just those for which at least one m_j is even. In other words, the order of the centre (namely 1 or 2) is the greatest common divisor of $m_1 + 1, \ldots, m_n + 1$.

Let b_k denote the number of isometries in the group that are expressible as products of k (but no fewer) reflections. Such an isometry leaves totally invariant an $(n - k)$-space, namely the intersection of k mirrors. Since the identity arises when $k = 0$, we have $b_0 = 1$. According to a remarkable theorem of SHEPHARD and TODD (1954, pp. 283, 290 to 294), these numbers can be identified with the elementary symmetric functions of the n m's; thus

$$\sum_{k=0}^{n} b_k t^k = \prod_{j=1}^{n} (1 + m_j t).$$

Setting $t = 1$, we find that the product $\Pi\,(m_j + 1)$ is equal to the order of the group.

For instance, the extended icosahedral group [3, 5] has exponents 1, 5, 9 and its 120 elements consist of E, 15 reflections (in planes through the pairs of opposite edges of the icosahedron), 59 rotations (namely, all of $[3, 5]^+$ except E) and 45 rotatory-reflections (namely Z, 20 of period six, and 24 of period ten), in agreement with the identity

$$(1 + t)(1 + 5t)(1 + 9t) = 1 + 15t + 59t^2 + 45t^3.$$

Since the characteristic roots ε^{m_j} occur in complex conjugate pairs, we have

$$m_1 + m_n = m_2 + m_{n-1} = \cdots = h$$

and $b_1 = \sum m_j = nh/2$. This simple expression for the number of reflections in the group was established by STEINBERG (1959, p. 500; see also COXETER 1963a, pp. 227–231).

The quadratic form 9.36 is the first of a set of n basic invariants $I_1, I_2, \ldots, I_n$, which generate the ring of polynomial invariants of the group (CHEVALLEY 1955a). It is remarkable that the degrees of these basic invariants are

$$m_1 + 1,\ m_2 + 1, \ldots, m_n + 1$$

(COLEMAN 1958, pp. 353–354; KOSTANT 1959, p. 1021; SOLOMON 1963). Moreover, the Jacobian

$$\frac{\partial(I_1, I_2, \ldots, I_n)}{\partial(x_1, x_2, \ldots, x_n)},$$

of order $\sum m_j = b_1$, factorizes into this number of linear forms which, when equated to zero, give the reflecting hyperplanes (COXETER 1951b, p. 775; SHEPHARD 1956, p. 47; STEINBERG 1960, p. 617).

KLEIN (1884, pp. 211—219) obtained invariants A, B, C, of degrees 2, 6, 10, for [3, 5]. Their Jacobian D is the product of 15 linear forms. Since each reflection reverses its sign, D is an invariant for $[3, 5]^+$, though not for [3, 5]; but

$$D^2 = -1728B^5 + C^3 + 720ACB^3 - 80A^2C^2B + 64A^3(5B^2 - AC)^2$$

is an invariant for [3, 5].

The *crystallographic* groups (as we remarked in § 9.3, p. 121) are those for which every p_{ik} is 1, 2, 3, 4 or 6. Another way of expressing this is that the number

$$f = \det(2a_{ik}) = 2^n \det(a_{ik})$$

is an integer (COXETER 1963a, pp. 206, 212; 1951a, pp. 414, 427). For each irreducible crystallographic group there is a family of simple groups of order

$$\frac{1}{d} \prod_{j=1}^{n} (q^{h+1} - q^{m_j}) = \frac{q^{hn/2}}{d} \prod_{k=1}^{n} (q^{m_k+1} - 1)$$

(CHEVALLEY 1955b, p. 64), where q is any power of any prime and $d = (f, q-1)$ except that, in the case of $[3^{n-3,1,1}]$, $d = (f, q^n - 1)$. The last column of Table 10 shows these groups in the notations of DICKSON (1901a, b) and ARTIN (1955, pp. 458—459). It is a strange thought that the order of $E_8(2)$, the first group in the last family, is comparable to EDDINGTON's estimate of the number of protons in the universe.

9.8 Infinite Euclidean groups. The coordinates $x_1, \ldots, x_n$ of 9.35 could more properly be written in the contravariant notation $x^1, \ldots, x^n$, so that 9.36 becomes

$$\sum \sum a_{ij} x^i x^j. \qquad (9.81)$$

In terms of the covariant coordinates

$$x_i = \sum a_{ij} x^j$$

the reflection R_k is expressed by

$$x_i' = x_i - 2a_{ik}x_k \qquad (i = 1, \ldots, n) \qquad (9.82)$$

(COXETER 1963a, p. 182). This leaves invariant the adjoint form

$$\sum \sum a^{ij} x_i x_j, \qquad (9.83)$$

where a^{ij} is the cofactor of a_{ij} in $\det(a_{ij})$. When 9.81 (and therefore also 9.83) is definite, x_k is simply the distance from $(x_1, \ldots, x_n)$ to the kth coordinate hyperplane $x_k = 0$, which contains all the x^i-axes except the x^k-axis.

It is remarkable that the affine reflections 9.82 ($k = 1, \ldots, n$) still generate the group 9.11 when 9.81 is only *semidefinite*, so that the group is infinite. In fact, since $\det(a_{ij}) = 0$, there exist constants $z^1, \ldots, z^n$ such that

$$\sum z^i a_{ik} = 0$$

and therefore, from 9.82,

$$\sum z^i x_i' = \sum z^i x_i. \tag{9.84}$$

These constants can be identified with the square roots of the coefficients of x_i^2 in the adjoint form 9.83, namely

$$z^i = \sqrt{a^{ii}}$$

(Coxeter 1963a, p. 177). By virtue of 9.84, we can regard the group as operating in the hyperplane

$$\sum z^i x_i = 1,$$

in which the sections of the coordinate hyperplanes $x_k = 0$ form an $(n-1)$-dimensional simplex. In this affine $(n-1)$-space, the semidefinite form determines a Euclidean metric, so that the R_k's appear as ordinary reflections in the bounding hyperplanes of this simplex (although in the whole n-space they can only be called "affine" reflections). Since the Euclidean $(n-1)$-space is simply-connected, the simplex serves as a fundamental region, and the reflections generate the infinite group 9.11 (Coxeter 1963a, pp. 80, 188).

The irreducible groups of this kind are listed in Table 11, where the third column gives the fundamental region in the notation of Coxeter (1963a, p. 194) and the fourth column gives the corresponding Lie group in the notation of Cartan (1927, pp. 218—225). This remarkable correspondence with simple Lie groups (or more precisely, with families of locally isomorphic simple Lie groups) was suggested by Cartan and rigorously established by Stiefel (1942; see also Coxeter 1951a, pp. 412, 426; Tits 1962, p. 205).

9.9 Infinite non-Euclidean groups. Lannér (1950, p. 53) has enumerated all irreducible groups 9.11 having the property that each subgroup generated by $n-1$ of the n R_k's is finite. In addition to the groups described above, he finds two infinite families with three generators, nine groups with four generators, and five with five generators (his Tables III and V). They include

$$[p, q], \quad [3, 5, 3], \quad [4, 3, 5], \quad [5, 3, 5],$$

$$[3, 3, 3, 5], \quad [4, 3, 3, 5], \quad [5, 3, 3, 5],$$

which are the symmetry groups of the regular hyperbolic honeycombs of Schlegel (1883, pp. 444, 454).

In all these cases the forms 9.81 and 9.83 are indefinite, and the x_i's can be regarded as homogeneous coordinates in a projective $(n-1)$-space in which the quadric

$$\sum \sum a^{ij} x_i x_j = 0$$

determines a hyperbolic metric.

When regarded as a projective collineation, 9.82 is a harmonic homology whose axial hyperplane is $x_k = 0$ while its centre is

$$(a_{1k}, \ldots, a_{nk}).$$

Since the centre and axial hyperplane are pole and polar, the hyperbolic metric makes this the reflection in $x_k = 0$. Thus all the R_k's are reflections in the bounding hyperplanes of the simplex of reference. Since the hyperbolic $(n-1)$-space (i.e., the interior of the absolute quadric) is simply-connected, this simplex serves as a fundamental region and the reflections generate the infinite group 9.11.

The same conclusion can be made for the still wider family of groups in which some or all of the subgroups generated by $n-1$ of the R_k's are not finite but "Euclidean". This merely means that the simplex has some or all of its vertices at infinity. Many such groups arise as groups of automorphs of indefinite quadratic forms; for instance, [4, 4, 3] is the group of automorphs of the quaternary form

$$x_0^2 - x_1^2 - x_2^2 - x_3^2,$$

which plays an important role in the special theory of relativity (COXETER and WHITROW 1950, p. 429), the automorphs being Lorentz transformations.

It is a natural conjecture that the affine reflections 9.35 or 9.82 will generate the group 9.11 for all values of the periods p_{ik} (with $p_{ii} = 1$). For instance, the group $[\infty]$, defined by

$$R_1^2 = R_2^2 = E,$$

is generated by the affine reflections 9.31 and 9.32 with $a = -1$, since their product

$$x_1' = -x_1 + 2x_2, \quad x_2' = -2x_1 + 3x_2$$

is a shear (VEBLEN and YOUNG 1918, p. 112; COXETER 1955, p. 12). Some progress in this direction has been made by BUNDGAARD (1952).

Many of the results in this chapter can be extended to groups generated by unitary reflections (§ 6.7, p. 79; see also SHEPHARD 1953; SHEPHARD and TODD 1954; COXETER 1957b). These groups are characterized by the following property. Whereas any group of linear transformations in n complex variables has a basic system of $n+1$ invariants, n of which are algebraically independent (BURNSIDE 1911, pp. 357—359), a group generated by reflections has a basic system of just n invariants (TODD 1947; 1950).

Tables 1—12

Table 1. *Non-Abelian groups of order less than 32* (§ 1.9)

Order	Symbol	Description	Abstract definition
6	$\mathfrak{D}_3 \cong \mathfrak{S}_3$	K-metacyclic	$S^3 = T^2 = (ST)^2 = E$
8	$\mathfrak{D}_4$	dihedral	$S^4 = T^2 = (ST)^2 = E$
	$\mathfrak{Q} \cong \langle 2, 2, 2\rangle$	quaternion	$S^2 = T^2 = (ST)^2$
10	$\mathfrak{D}_5$	ZS-metacyclic	$S^5 = T^2 = (ST)^2 = E$
12	$\mathfrak{D}_6 \cong \mathfrak{C}_2 \times \mathfrak{D}_3$	dihedral	$S^6 = T^2 = (ST)^2 = E$
	$\mathfrak{A}_4$	tetrahedral	$S^3 = T^2 = (ST)^3 = E$
	$\langle 2, 2, 3\rangle$	ZS-metacyclic	$S^3 = T^2 = (ST)^2$
14	$\mathfrak{D}_7$	ZS-metacyclic	$S^7 = T^2 = (ST)^2 = E$
16	$\mathfrak{C}_2 \times \mathfrak{D}_4$		$R_1^2 = R_2^2 = R_3^2 = (R_3 R_1)^2$ $= (R_1 R_2)^2 = (R_2 R_3)^4 = E$
	$\mathfrak{C}_2 \times \mathfrak{Q}$		$R^2 = S^2 = (RS)^2$, $T^2 = E$, $T \rightleftarrows R, S$
	$\mathfrak{D}_8$	dihedral	$S^8 = T^2 = (ST)^2 = E$
	$\langle -2, 4 \mid 2\rangle$		$\begin{cases} T^2 = E,\ TST = S^3 \text{ or} \\ R^2 S^4 = (RS)^2 = E \end{cases}$
	$\langle 2, 2 \mid 2\rangle$		$\begin{cases} T^2 = E,\ TST = S^{-3} \text{ or} \\ R^2 = S^2,\ (RS)^2 = E \end{cases}$
	$\langle 2, 2 \mid 4; 2\rangle$		$S^4 = T^4 = E$, $T^{-1} S T = S^{-1}$
	$(4, 4 \mid 2, 2)$		$R^4 = S^4 = (RS)^2$ $= (R^{-1} S)^2 = E$
	$\langle 2, 2, 2\rangle_2$	Figs. 3.3c, 3.6c	$R^2 = S^2 = T^2 = E$, $RST = STR = TRS$
	$\langle 2, 2, 4\rangle$	dicyclic	$S^4 = T^2 = (ST)^2$
18	$\mathfrak{C}_3 \times \mathfrak{D}_3$	Z-metacyclic	$\begin{cases} R^2 = S^3 = E, (RS)^2 = (SR)^2 \text{ or} \\ S^3 = T^6 = E,\ T^{-1} S T = S^{-1} \end{cases}$
	$\mathfrak{D}_9$	ZS-metacyclic	$S^9 = T^2 = (ST)^2 = E$
	$((3, 3, 3; 2))$		$R^2 = S^2 = T^2 = (RST)^2$ $= (RS)^3 = (RT)^3 = E$
20	$\mathfrak{D}_{10} \cong \mathfrak{C}_2 \times \mathfrak{D}_5$	dihedral	$S^{10} = T^2 = (ST)^2 = E$
	$F^{2,1,-1}$	K-metacyclic	$S^2 TSTS^{-1} T = T^2 = E$
	$\langle 2, 2, 5\rangle$	ZS-metacyclic	$S^5 = T^2 = (ST)^2$
21		ZS-metacyclic	$T^3 = E$, $T^{-1} S T = S^2$
22	$\mathfrak{D}_{11}$	ZS-metacyclic	$S^{11} = T^2 = (ST)^2 = E$
24	$\mathfrak{C}_2 \times \mathfrak{A}_4$		$S^3 = T^2 = (S^{-1} TST)^2 = E$
	$\mathfrak{C}_2 \times \mathfrak{D}_6 \cong \mathfrak{D}_2 \times \mathfrak{D}_3$		$R_1^2 = R_2^2 = R_3^2 = (R_2 R_3)^6$ $= (R_3 R_1)^2 = (R_1 R_2)^2 = E$
	$\mathfrak{C}_3 \times \mathfrak{D}_4$		$S^{12} = T^2 = E$, $TST = S^{-5}$
	$\mathfrak{C}_3 \times \mathfrak{Q}$		$S^{12} = E$, $T^2 = S^6$, $T^{-1} S T = S^7$
	$\mathfrak{C}_4 \times \mathfrak{D}_3$		$S^{12} = T^2 = E$, $TST = S^5$
	$\mathfrak{C}_2 \times \langle 2, 2, 3\rangle$		$S^6 = T^4 = E$, $T^{-1} S T = S^{-1}$
	$\mathfrak{D}_{12}$	dihedral	$S^{12} = T^2 = (ST)^2 = E$
	$\mathfrak{S}_4$	octahedral	$S^4 = T^2 = (ST)^3 = E$
	$\langle 2, 3, 3\rangle$	binary tetrahedral	$\begin{cases} R^3 = S^3 = (RS)^2 \text{ or} \\ A^3 = E,\ ABA = BAB \text{ or} \\ S^3 = T^2,\ (S^{-1} T)^3 = E \end{cases}$

Table 1 (Continued)

Order	Symbol	Description	Abstract definition
	$(4, 6 \mid 2, 2)$	Fig. 8.5	$R^4 = S^6 = (RS)^2 = (R^{-1}S)^2 = E$
	$\langle -2, 2, 3 \rangle$	ZS-metacyclic	$S^2 = T^2 = (ST)^3$
	$\langle 2, 2, 6 \rangle$	dicyclic	$S^6 = T^2 = (ST)^2$
26	$\mathfrak{D}_{13}$	ZS-metacyclic	$S^{13} = T^2 = (ST)^2 = E$
27	$(3, 3 \mid 3, 3)$		$R^3 = S^3 = (RS)^3 = (R^{-1}S)^3 = E$
			$T^3 = E,\ T^{-1}ST = S^{-2}$
28	$\mathfrak{D}_{14} \cong \mathfrak{C}_2 \times \mathfrak{D}_7$	dihedral	$S^{14} = T^2 = (ST)^2 = E$
	$\langle 2, 2, 7 \rangle$	ZS-metacyclic	$S^7 = T^2 = (ST)^2$
30	$\mathfrak{C}_3 \times \mathfrak{D}_5$	ZS-metacyclic	$T^2 = E,\ TST = S^4$ or $s^5 = t^6 = E,\ t^{-1}st = s^{-1}$
	$\mathfrak{C}_5 \times \mathfrak{D}_3$	ZS-metacyclic	$T^2 = E,\ TST = S^{-4}$ or $s^3 = t^{10} = E,\ t^{-1}st = s^{-1}$
	$\mathfrak{D}_{15}$	ZS-metacyclic	$S^{15} = T^2 = (ST)^2 = E$

Table 2. *The crystallographic and non-crystallographic point groups* (§§ 4.3, 4.4)

Hermann & Mauguin	Weyl	Pólya & Meyer	Schoenflies	Coxeter	Structure	Order	q
$\mathbf{q}$	C_q	C_q	C_q	$[q]^+$	$\mathfrak{C}_q$	q	1, 2, ...
$\bar{\mathbf{1}}$	$\bar{C}_1$	C_{1i}	$C_i = S_2$	$[2^+, 2^+]$	$\mathfrak{C}_2 \cong \mathfrak{D}_1$	2	
$\mathbf{m} = \bar{\mathbf{2}}$	$C_2 C_1$	$\mathsf{C}_1 [\mathsf{C}_2$	$C_s = C_{1h} = C_{1v}$	$[1]$			
$\bar{\mathbf{q}}$	$\bar{C}_q$	C_{qi}	S_{2q}	$[2^+, 2q^+]$	$\mathfrak{C}_{2q} \cong \mathfrak{C}_2 \times \mathfrak{C}_q$	$2q$	odd
$\overline{\mathbf{2q}}$	$C_{2q} C_q$	$\mathsf{C}_q [\mathsf{C}_{2q}$	C_{qh}	$[2, q^+]$			
$\overline{\mathbf{2q}}$	$C_{2q} C_q$	$\mathsf{C}_q [\mathsf{C}_{2q}$	S_{2q}	$[2^+, 2q^+]$	$\mathfrak{C}_{2q}$	$2q$	even
$\mathbf{q/m}$	$\bar{C}_q$	C_{qi}	C_{qh}	$[2, q^+]$	$\mathfrak{C}_2 \times \mathfrak{C}_q$		
$\mathbf{2\,2\,2}$	D_2	D_2	$V = D_2$	$[2, 2]^+$	$\mathfrak{D}_2 \cong \mathfrak{C}_2 \times \mathfrak{C}_2$	4	
$\mathbf{m\,m\,2}$	$D_2 C_2$	$\mathsf{C}_2 [\mathsf{D}_2$	C_{2v}	$[2]$			
$\mathbf{q\,2}$	D_q	D_q	D_q	$[2, q]^+$	$\mathfrak{D}_q$	$2q$	2, 3, ...
$\mathbf{q\,m}$	$D_q C_q$	$\mathsf{C}_q [\mathsf{D}_q$	C_{qv}	$[q]$			
$\bar{\mathbf{4}}\,\mathbf{2\,m}$	$D_4 D_2$	$\mathsf{D}_2 [\mathsf{D}_4$	$V_d = D_{2d}$	$[2^+, 4]$	$\mathfrak{D}_4$	8	
$\overline{\mathbf{2q}}\,\mathbf{2\,m}$	$D_{2q} D_q$	$\mathsf{D}_q [\mathsf{D}_{2q}$	D_{qd}	$[2^+, 2q]$	$\mathfrak{D}_{2q}$	$4q$	even
$\bar{\mathbf{q}}\,\mathbf{m}$	$\bar{D}_q$	D_{qi}	D_{qd}	$[2^+, 2q]$	$\mathfrak{D}_{2q} \cong \mathfrak{C}_2 \times \mathfrak{D}_q$	$4q$	odd
$\overline{\mathbf{2q}}\,\mathbf{m\,2}$	$D_{2q} D_q$	$\mathsf{D}_q [\mathsf{D}_{2q}$	D_{qh}	$[2, q]$			
$\mathbf{m\,m\,m}$	$\bar{D}_2$	D_{2i}	$V_h = D_{2h}$	$[2, 2]$	$\mathfrak{C}_2 \times \mathfrak{C}_2 \times \mathfrak{C}_2$	8	
$\mathbf{q/m\,m\,m}$	$\bar{D}_q$	D_{qi}	D_{qh}	$[2, q]$	$\mathfrak{C}_2 \times \mathfrak{D}_q$	$4q$	even
$\mathbf{2\,3}$	T	T	T	$[3, 3]^+$	$\mathfrak{A}_4$	12	
$\mathbf{m\,3}$	$\bar{T}$	T_i	T_h	$[3^+, 4]$	$\mathfrak{C}_2 \times \mathfrak{A}_4$	24	
$\mathbf{4\,3\,2}$	W	O	O	$[3, 4]^+$	$\mathfrak{S}_4$	24	
$\bar{\mathbf{4}}\,\mathbf{3\,m}$	WT	$\mathsf{T} [\mathsf{O}$	T_d	$[3, 3]$			
$\mathbf{m\,3\,m}$	$\bar{W}$	O_i	O_h	$[3, 4]$	$\mathfrak{C}_2 \times \mathfrak{S}_4$	48	
$\mathbf{5\,3\,2}$	P	I	I	$[3, 5]^+$	$\mathfrak{A}_5$	60	
$\mathbf{5\,3\,m}$	$\bar{P}$	I_i	I_h	$[3, 5]$	$\mathfrak{C}_2 \times \mathfrak{A}_5$	120	

Table 3. *The 17 space groups of 2-dimensional crystallography* (§ 4.5)

Hermann & Mauguin	Pólya	Niggli	Speiser (Abb.)	Fricke & Klein	Fig. in this book	Abstract definition
p 1	C_1	$\mathfrak{C}_1{}^{\mathrm{I}}$	17		4.5a, b	$XY = YX$, or $XYZ = ZYX = E$
p 2	C_2	$\mathfrak{C}_2{}^{\mathrm{I}}$	18	65	4.5c, d	$T_i^2 = (T_1T_2T_3)^2 = E$, or $T_j^2 = T_1T_2T_3T_4 = E$
p m	D_1kk	$\mathfrak{C}_s{}^{\mathrm{I}}$	19	62	4.5e	$R^2 = R'^2 = E,\ RY = YR,\ R'Y = YR'$
p g	D_1gg	$\mathfrak{C}_s{}^{\mathrm{II}}$	20	63	4.5f	$P^2 = Q^2$
c m	D_1kg	$\mathfrak{C}_s{}^{\mathrm{III}}$	21	60	4.5g	$R^2 = E,\ RP^2 = P^2R$
p m m	D_2kkkk	$\mathfrak{C}_{2v}{}^{\mathrm{I}}$	22	65	4.5h	$R_j^2 = (R_1R_2)^2 = (R_2R_3)^2 = (R_3R_4)^2 = (R_4R_1)^2 = E$
p m g	D_2kkgg	$\mathfrak{C}_{2v}{}^{\mathrm{III}}$	24	66	4.5i	$R^2 = T_1^2 = T_2^2 = E,\ T_1RT_1 = T_2RT_2$
p g g	D_2gggg	$\mathfrak{C}_{2v}{}^{\mathrm{II}}$	23	67	4.5j	$(PO)^2 = (P^{-1}O)^2 = E$
c m m	D_2kgkg	$\mathfrak{C}_{2v}{}^{\mathrm{IV}}$	25	64	4.5k	$R_1^2 = R_2^2 = T^2 = (R_1R_2)^2 = (R_1TR_2T)^2 = E$
p 4	C_4	$\mathfrak{C}_4{}^{\mathrm{I}}$	26	71	4.5l	$S^4 = T^2 = (ST)^4 = E$
p 4 m	D_4^*	$\mathfrak{C}_{4v}{}^{\mathrm{I}}$	27	71	4.5m	$R_i^2 = (R_1R_2)^4 = (R_2R_3)^4 = (R_3R_1)^2 = E$
p 4 g	D_4^0	$\mathfrak{C}_{4v}{}^{\mathrm{II}}$	28	72	4.5n	$S^4 = R^2 = (S^{-1}RSR)^2 = E$
p 3	C_3	$\mathfrak{C}_3{}^{\mathrm{I}}$	29	68	4.5o, p	$S_1^3 = S_2^3 = (S_1S_2)^3 = E$, or $S_i^3 = S_1S_2S_3 = E$
p 3 1 m	D_3^0	$\mathfrak{C}_{3v}{}^{\mathrm{II}}$	30	69	4.5q	$S^3 = R^2 = (S^{-1}RSR)^3 = E$
p 3 m 1	D_3^*	$\mathfrak{C}_{3v}{}^{\mathrm{I}}$	31	68	4.5r	$R_i^2 = (R_1R_2)^3 = (R_2R_3)^3 = (R_3R_1)^3 = E$
p 6	C_6	$\mathfrak{C}_6{}^{\mathrm{I}}$	32	70	4.5s	$S^3 = T^2 = (ST)^6 = E$
p 6 m	D_6	$\mathfrak{C}_{6v}{}^{\mathrm{I}}$	33	70	4.5t	$R_i^2 = (R_1R_2)^3 = (R_2R_3)^6 = (R_3R_1)^2 = E$

Table 4. *Subgroup relationships among the 17 plane crystallographic groups* (§ 4.6)

	p1	**p2**	**pg**	**pm**	**cm**	**pgg**	**pmg**	**pmm**	**cmm**	**p4**	**p4g**	**p4m**	**p3**	**p31m**	**p3m1**	**p6**	**p6m**
p1	2																
p2	2	2															
pg	2		2														
pm	2		2	2	2												
cm	2		2	2	*3*												
pgg	4	2	2			*3*											
pmg	4	2	2	2	*4*	2	*3*										
pmm	4	2	4	2	4	4	2	2	2								
cmm	4	2	4	4	2	2	2	2	*4*								
p4	4	2								2							
p4g	8	4	*4*	*8*	*4*	2	*4*	4	2	2	*9*						
p4m	8	4	*8*	*4*	*4*	4	*4*	2	2	2	2	2					
p3	3												3				
p31m	6		*6*	*6*	*3*								2	*4*	3		
p3m1	6		*6*	*6*	*3*								2	*3*	*4*		
p6	6	3											2			*4*	
p6m	12	6	*12*	*12*	*6*	*6*	*6*	*6*	*3*				4	2	2	2	*3*

Table 5. *Alternating and symmetric groups of degree less than 8* (§ 6.4)

Group	Abstract definition	Generators	References
$\mathfrak{A}_3$	$S^3 = E$	$S = (1\ 2\ 3)$	
$\mathfrak{S}_3$	$R^2 = S^3 = (RS)^2 = E$	$R = (1\ 2)$, $S = (1\ 2\ 3)$	
$\mathfrak{A}_4$	$R^2 = S^3 = (RS)^3 = E$	$R = (1\ 2)\ (3\ 4)$, $S = (2\ 3\ 4)$	
$\mathfrak{S}_4$	$R^2 = S^3 = (RS)^4 = E$		Dyck 1882, p. 35
	$R^2 = S^3 = T^4 = RST = E$	$R = (3\ 4)$, $S = (1\ 2\ 3)$, $T = (4\ 3\ 2\ 1)$	
	$S^3 = T^4 = (ST)^2 = E$		
$\mathfrak{A}_5$	$R^2 = S^3 = (RS)^5 = E$	$R = (1\ 2)\ (4\ 5)$, $S = (1\ 3\ 4)$	Hamilton 1856, p. 446
	$B^5 = C^5 = (BC)^2 = (B^{-1}C)^3 = E$	$A = (5\ 3\ 2)$,	Todd and Coxeter 1936, p. 31
	$B^5 = (B^2D)^2 = E$, $BDB = DBD$	$B = (1\ 2\ 3\ 4\ 5)$, $C = (1\ 2\ 4\ 3\ 5)$	
	$A^3 = B^5 = C^5 = (AB)^2 = (BC)^2 = (CA)^2 = (ABC)^2 = E$	$D = (1\ 4\ 3\ 5\ 2)$	Coxeter 1939, p. 107
	$A^2 = B^2 = C^2 = (AB)^3 = (BC)^5 = (CA)^3 = (ABC)^3 = E$	$A = (1\ 2)\ (4\ 5)$, $B = (1\ 2)\ (3\ 4)$, $C = (2\ 3)\ (4\ 5)$	Coxeter 1939, p. 144
	$V_1^3 = V_2^3 = V_3^3 = (V_1V_2)^2 = (V_2V_3)^2 = (V_3V_1)^2 = E$	$V_1 = (1\ 2\ 5)$, $V_2 = (1\ 3\ 5)$, $V_3 = (1\ 4\ 5)$	Coxeter 1934c, p. 218
$\mathfrak{S}_5$	$A^2 = B^5 = (AB)^4 = (AB^{-1}AB)^3 = E$	$A = (1\ 2)$, $B = (1\ 2\ 3\ 4\ 5)$	Burnside 1897, p. 125
	$A^2 = B^5 = (AB)^4 = (AB^{-2}AB^2)^2 = E$		Coxeter and Todd 1936, p. 197
	$R^5 = S^4 = (RS)^2 = (R^2S^2)^3 = E$	$R = (5\ 4\ 3\ 2\ 1)$,	
	$S^4 = T^6 = (ST)^2 = (S^{-1}T)^3 = E$	$S = (1\ 2\ 3\ 4)$,	Coxeter and Todd 1936, p. 195
	$R^5 = T^6 = (RT)^2 = (R^2T^2)^2 = E$	$T = (1\ 2)\ (3\ 4\ 5)$	
$\mathfrak{A}_6$	$A^5 = B^5 = (AB)^2 = (A^{-1}B)^4 = E$	$A = (1\ 6\ 5\ 4\ 3)$, $B = (1\ 2\ 3\ 4\ 5)$	Coxeter 1939, pp. 84, 91
	$A^3 = B^4 = (AB)^5 = (A^{-1}B^{-1}AB)^2 = E$	$A = (1\,3\,5)\,(2\,4\,6)$, $B = (1\,4)\,(2\,6\,3\,5)$	Coxeter 1939, p. 103
	$L^2 = M^2 = N^2 = (LM)^3 = (MN)^3 = (LN)^4 = (LMN)^5 = E$	$L = (2\ 3)\ (4\ 6)$, $M = (1\ 2)\ (4\ 5)$, $N = (1\ 2)\ (3\ 5)$	Coxeter 1970, pp. 28, 33, 34, 46.
$\mathfrak{S}_6$	$A^2 = (AB^{-1}AB)^3 = (AB^{-2}AB^2)^2 = E$, $B^6 = (AB)^5$	$A = (1\ 2)$, $B = (1\,2\,3\,4\,5\,6)$	(6.271)
$\mathfrak{A}_7$	$A^4 = B^5 = (AB)^3 = (A^{-1}BA^2B^2)^2 = E$	$A = (1\,2)\ (3\,4\,5\,6)$, $B = (2\ 3\ 7\ 5\ 4)$	K. C. Young
$\mathfrak{S}_7$	$A^2 = (AB^{-2}AB^2)^2 = (AB^{-3}AB^3)^2 = E$, $B^7 = (AB)^6$	$A = (1\ 2)$, $B = (1\,2\,3\,4\,5\,6\,7)$	(6.27)

Table 6. *The groups $LF(2, p)$ for $2 < p < 30$ (§ 7.5)*

p	Order $p(p^2-1)/2$	Generators (mod p)	Abstract definition
3	12	$S = \begin{pmatrix} 1 & 0 \\ 1 & 1 \end{pmatrix}, T = \begin{pmatrix} 0 & 1 \\ -1 & 0 \end{pmatrix}$	$S^3 = T^2 = (ST)^3 = E$
5	60	$S = \begin{pmatrix} 1 & 0 \\ 1 & 1 \end{pmatrix}, T = \begin{pmatrix} 0 & 1 \\ -1 & 0 \end{pmatrix}$	$S^5 = T^2 = (ST)^3 = E$
7	168	$R = \begin{pmatrix} 0 & 1 \\ -1 & -1 \end{pmatrix}, S = \begin{pmatrix} 2 & 0 \\ 1 & -3 \end{pmatrix}$	$R^3 = S^3 = (RS)^4 = (R^{-1}S)^4 = E$
11	660	$S = \begin{pmatrix} 1 & 0 \\ 1 & 1 \end{pmatrix}, T = \begin{pmatrix} 0 & 1 \\ -1 & 0 \end{pmatrix}$	$S^{11} = T^2 = (ST)^3 = (S^4TS^6T)^2 = E$
13	1092	$S = \begin{pmatrix} 5 & 4 \\ -6 & -2 \end{pmatrix}, T = \begin{pmatrix} -5 & 0 \\ 0 & 5 \end{pmatrix}$	$S^7 = T^2 = (ST)^6 = (S^2T)^3 = E$
17	2448	$S = \begin{pmatrix} 5 & 8 \\ 5 & -2 \end{pmatrix}, T = \begin{pmatrix} -4 & 0 \\ 0 & 4 \end{pmatrix}$	$S^9 = T^2 = (ST)^4 = (S^2T)^3 = E$
19	3420	$S = \begin{pmatrix} 4 & 0 \\ 0 & 5 \end{pmatrix}, T = \begin{pmatrix} 4 & 4 \\ -9 & -4 \end{pmatrix}$	$S^9 = T^2 = (ST)^5 = (S^{-1}TST)^2 = E$
23	6072	$S = \begin{pmatrix} 3 & 0 \\ 0 & 8 \end{pmatrix}, T = \begin{pmatrix} 9 & 9 \\ -4 & -9 \end{pmatrix}$	$S^{11} = T^2 = (ST)^3 = (S^{-1}TST)^4 = E$
29	12180	$P = \begin{pmatrix} -7 & 12 \\ 11 & 10 \end{pmatrix}, Q = \begin{pmatrix} 4 & -8 \\ 9 & 4 \end{pmatrix}$	$P^7 = (P^2Q)^3 = (P^3Q)^2 = (PQ^8)^2 = E$

Table 7. *The simplest reflexible maps* (§§ 8.3–8.6)

Characteristic	Genus (when orientable)	Map	Vertices	Edges	Faces	Group	Order
2	0	$\{p, 2\}$	p	p	2	$\mathfrak{C}_2 \times \mathfrak{D}_p$	$4p$
		$\{2, p\}$	2	p	p		
		$\{3, 3\}$	4	6	4	$\mathfrak{S}_4$	24
		$\{4, 3\}$	8	12	6	$\mathfrak{C}_2 \times \mathfrak{S}_4$	48
		$\{3, 4\}$	6	12	8		
		$\{5, 3\}$	20	30	12	$\mathfrak{C}_2 \times \mathfrak{A}_5$	120
		$\{3, 5\}$	12	30	20		
1	—	$\{2q, 2\}/2$	q	q	1	$\mathfrak{D}_{2q}$	$4q$
		$\{2, 2q\}/2$	1	q	q		
		$\{4, 3\}/2 = \{4, 3\}_3$	4	6	3	$\mathfrak{S}_4$	24
		$\{3, 4\}/2 = \{4, 3\}_3$	3	6	4		
		$\{5, 3\}/2 = \{5, 3\}_5$	10	15	6	$\mathfrak{A}_5$	60
		$\{3, 5\}/2 = \{3, 5\}_5$	6	15	10		
0	1	$\{4, 4\}_{q,0} = \{4, 4 \mid q\}$	q^2	$2q^2$	q^2	$(4, 4 \mid 2, q)$	$8q^2$
		$\{6, 3\}_{q,0} = \{6, 3\}_{2q}$	$2q^2$	$3q^2$	q^2	$G^{3,6,2q}$	$12q^2$
		$\{3, 6\}_{q,0} = \{3, 6\}_{2q}$	q^2	$3q^2$	$2q^2$		
		$\{4, 4\}_{q,q} = \{4, 4\}_{2q}$	$2q^2$	$4q^2$	$2q^2$	$G^{4,4,2q}$	$16q^2$
		$\{6, 3\}_{q,q}$	$6q^2$	$9q^2$	$3q^2$	8.43, 8.431	$36q^2$
		$\{3, 6\}_{q,q}$	$3q^2$	$9q^2$	$6q^2$		
−2	—	$\{6, 4\}_3$	6	12	4	$\mathfrak{C}_2 \times \mathfrak{S}_4$	48
		$\{4, 6\}_3$	4	12	6		

Table 8. *The known finite maps* $\{p, q\}_r$ (§ 8.6)
(To save space, only one of each pair of duals is listed: the one with $p \geqq q$)

Map	Vertices	Edges	Faces	Characteristic	Genus (when orientable)	Group	Order
$\{2q, 2\}_{2q}$	$2q$	$2q$	2	2	0	$\mathfrak{C}_2 \times \mathfrak{D}_{2q}$	$8q$
$\{2q, 2q\}_2$	2	$2q$	2	$4-2q$	$q-1$		
$\{2p+1, 2\}_{4p+2}$	$2p+1$	$2p+1$	2	2	0	$\mathfrak{D}_{4p+2}$	$8p+4$
$\{4p+2, 2\}_{2p+1}$	$2p+1$	$2p+1$	1	1	—		
$\{4p+2, 2p+1\}_2$	2	$2p+1$	1	$2-2p$	p		
$\{6, 3\}_{2q}$	$2q^2$	$3q^2$	q^2	0	1	$G^{3,6,2q}$ ($\cong \mathfrak{C}_2 \times \mathfrak{S}_{q+1}$ when $q = 1$ or 2)	$12q^2$
$\{2q, 3\}_6$	$2q^2$	$3q^2$	$3q$	$(3-q)q$	$\binom{q-1}{2}$		
$\{2q, 6\}_3$	q^2	$3q^2$	$3q$	$(3-2q)q$	—		
$\{4, 4\}_{2q}$	$2q^2$	$4q^2$	$2q^2$	0	1	$G^{4,4,2q}$	$16q^2$
$\{2q, 4\}_4$	$2q^2$	$4q^2$	$4q$	$(4-2q)q$	$(q-1)^2$		
$\{3, 3\}_4$	4	6	4	2	0	$\mathfrak{S}_4$	24
$\{4, 3\}_3$	4	6	3	1	—		
$\{5, 3\}_5$	10	15	6	1	—	$\mathfrak{A}_5$	60
$\{5, 5\}_3$	6	15	6	-3	—		
$\{5, 3\}_{10}$	20	30	12	2	0	$\mathfrak{C}_2 \times \mathfrak{A}_5$	120
$\{10, 3\}_5$	20	30	6	-4	—		
$\{10, 5\}_3$	12	30	6	-12	—		
$\{5, 4\}_5$	20	40	16	-4	—	$(4, 5 \mid 2, 4)$ (Coxeter 1937a, p. 53)	160
$\{5, 5\}_4$	16	40	16	-8	5		
$\{5, 4\}_6$	30	60	24	-6	4	$\mathfrak{C}_2 \times \mathfrak{S}_5$	240
$\{6, 4\}_5$	30	60	20	-10	—		
$\{6, 5\}_4$	24	60	20	-16	9		
$\{7, 3\}_8$	56	84	24	-4	3	$PGL(2, 7)$	336
$\{8, 3\}_7$	56	84	21	-7	—		
$\{8, 7\}_3$	24	84	21	-39	—		
$\{7, 3\}_9$	84	126	36	-6	—	$LF(2, 2^3)$	504
$\{9, 3\}_7$	84	126	28	-14	—		
$\{9, 7\}_3$	36	126	28	-62	—		
$\{5, 5\}_5$	66	165	66	-33	—	$LF(2, 11)$	660
$\{8, 3\}_8$	112	168	42	-14	8	$\mathfrak{C}_2 \times PGL(2, 7)$	672
$\{8, 8\}_3$	42	168	42	-84	—		
$\{5, 4\}_8$	180	360	144	-36	19	$\mathfrak{C}_2 \times PGL(2, 3^2)$	1440
$\{8, 4\}_5$	180	360	90	-90	—		
$\{8, 5\}_4$	144	360	90	-126	64		
$\{7, 3\}_{12}$	364	546	156	-26	14	$PGL(2, 13)$	2184
$\{12, 3\}_7$	364	546	91	-91	—		
$\{12, 7\}_3$	156	546	91	-299	—		

Table 8 (Continued)

Map	Vertices	Edges	Faces	Characteristic	Genus (when orientable)	Group	Order
$\{7, 3\}_{13}$	182	273	78	−13	—	$LF(2, 13)$	1092
$\{13, 3\}_7$	182	273	42	−49	—		
$\{13, 7\}_3$	78	273	42	−153	—		
$\{7, 3\}_{14}$	364	546	156	−26	—	$PGL(2, 13)$	2184
$\{14, 3\}_7$	364	546	78	−104	14		
$\{14, 7\}_3$	156	546	78	−312	—		
$\{9, 3\}_9$	570	855	190	−95	—	$LF(2, 19)$	3420
$\{9, 9\}_3$	190	855	190	−475	—		
$\{8, 3\}_{10}$	720	1080	270	−90	46	$\mathfrak{C}_2 \times (2, 3, 8; 5)$ (Coxeter 1939, p. 92)	4320
$\{10, 3\}_8$	720	1080	216	−144	73		
$\{10, 8\}_3$	270	1080	216	−594	—		
$\{6, 4\}_7$	546	1092	364	−182	—	$\mathfrak{C}_2 \times PGL(2, 13)$	4368
$\{7, 4\}_6$	546	1092	312	−234	118		
$\{7, 6\}_4$	364	1092	312	−416	209		
$\{5, 4\}_9$	855	1710	684	−171	—	$PGL(2, 19)$	6840
$\{9, 4\}_5$	855	1710	380	−475	—		
$\{9, 5\}_4$	684	1710	380	−646	324		
$\{8, 3\}_{11}$	2024	3036	759	−253	—	$PGL(2, 23)$	12144
$\{11, 3\}_8$	2024	3036	552	−460	231		
$\{11, 8\}_3$	759	3036	552	−1725	—		
$\{7, 3\}_{15}$	2030	3045	870	−145	—	$LF(2, 29)$	12180
$\{15, 3\}_7$	2030	3045	406	−609	—		
$\{15, 7\}_3$	870	3045	406	−1769	—		
$\{9, 3\}_{10}$	3420	5130	1140	−570	286	$\mathfrak{D}_3 \times LF(2, 19)$	20520
$\{10, 3\}_9$	3420	5130	1026	−684	—		
$\{10, 9\}_3$	1140	5130	1026	−2964	—		
$\{7, 3\}_{16}$	3584	5376	1536	−256	129	$G^{3,7,16}$	21504
$\{16, 3\}_7$	3584	5376	672	−1120	—		
$\{16, 7\}_3$	1536	5376	672	−3168	—		

Table 9. *The regular maps of genus 2* (§§ 8.5—8.8)

Map	Vertices	Edges	Faces	"Rotation" group	Order	Brahana (1927, p. 284)	Threlfall (1932a, p. 44)	Bilinski (1950, p. 148)
$\{8, 8\}_{1,0}$	1	4	1	$\mathfrak{C}_8$	8	1 octagon	14	XIX
$\{10, 5\}_2$	2	5	1	$\mathfrak{C}_{10}$	10	1 decagon	12	XXII
$\{5, 10\}_2$	1	5	2			2 pentagons		XIII
$\{6, 6\}_2$	2	6	2	$\mathfrak{C}_2 \times \mathfrak{C}_6$	12	2 hexagons	13	XV
$\{8, 4\}_{1,1}$	4	8	2	$\langle -2, 4 \mid 2\rangle$	16	2 octagons	9	XVIII
$\{4, 8\}_{1,1}$	2	8	4			4 quadrangles		IX
$\{6, 4 \mid 2\}$	6	12	4	$(4, 6 \mid 2, 2)$	24	—	8	XIV
$\{4, 6 \mid 2\}$	4	12	6					VIII
$\{4+4, 3\}$	16	24	6	$\langle -3, 4 \mid 2\rangle$	48	6 octagons	2	XVII
$\{3, 4+4\}$	6	24	16			16 triangles		II

Table 10. *The irreducible finite groups of the form* $(R_i R_k)^{p_{ik}} = E$, $p_{ii} = 1$ (see §§ 9.3, 9.7)

Graph	Number of generators (nodes)	Abbreviated symbol	Order	Order of Centre	$f=\det(2a_{ik})$	h	$m_1, m_2, \ldots, m_{n-1}, m_n$	Corresponding family of simple groups
	$n \geqq 1$	$[3^{n-1}]$	$(n+1)!$	1	$n+1$	$n+1$	$1, 2, \ldots, n-1, n$	$LF(n+1, q) \cong L_{n+1}(q)$
	$n \geqq 2$	$[3^{n-2}, 4]$	$2^n n!$	2	2	$2n$	$1, 3, \ldots, 2n-3, 2n-1$	$A(2n, q) \cong S_{2n}(q)$
	$n \geqq 4$	$[3^{n-3,1,1}]$	$2^{n-1} n!$	$(n, 2)$	4	$2n-2$	$1, 3, \ldots, n-1, \ldots, 2n-5, 2n-3$	$FO(2n, q) \cong O_{2n}(1, q)$
	2	$[r]$	$2r$	$(r, 2)$	$4 \sin^2 \pi/r$	r	$1, r-1$	$H' \cong G_2(q)$ $(r = 6)$
	3	$[3, 5]$	120	2	$2\tau^{-2}$	10	1, 5, 9	—
	4	$[3, 4, 3]$	1152	2	1	12	1, 5, 7, 11	$F_4(q)$
	4	$[3, 3, 5]$	14400	2	τ^{-4}	30	1, 11, 19, 29	—
	6	$[3^{2,2,1}]$	$51840 = 2^7 \cdot 3^4 \cdot 5$	1	3	12	1, 4, 5, 7, 8, 11	$E_6(q)$
	7	$[3^{3,2,1}]$	$2903040 = 2^{10} \cdot 3^4 \cdot 5 \cdot 7$	2	2	18	1, 5, 7, 9, 11, 13, 17	$E_7(q)$
	8	$[3^{4,2,1}]$	$696729600 = 2^{14} \cdot 3^5 \cdot 5^2 \cdot 7$	2	1	30	1, 7, 11, 13, 17, 19, 23, 29	$E_8(q)$

Table 11. *The irreducible Euclidean groups of the form* $(R_i R_k)^{p_{ik}} = E$ $(p_{ii} = 1)$ (§9.8)

Graph	Number of generators (nodes)	Fundamental region	Corresponding Simple Lie group
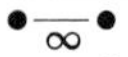 ∞	2	W_2	A_1
 …	$n \geqq 3$	P_n	A_{n-1}
 4	$n \geqq 4$	S_n	B_{n-1}
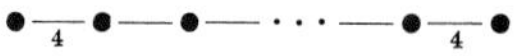 4 4	$n \geqq 3$	R_n	C_{n-1}
	$n \geqq 5$	Q_n	D_{n-1}
 6	3	V_3	G_2
 4	5	U_5	F_4
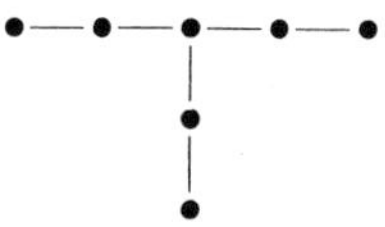	7	T_7	E_6
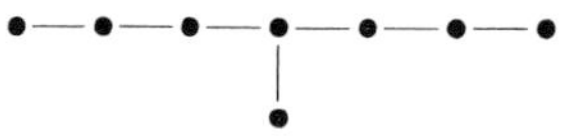	8	T_8	E_7
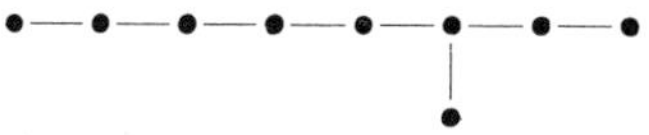	9	T_9	E_8

Table 12. *Glossary of symbols used for particular abstract definitions*

Symbol	Abstract definition	Reference
$\mathfrak{C}_n \cong [n]^+$	$R^n = E$	§§ 1.1, 9.5
$\mathfrak{D}_n \cong [n] \cong (2, 2, n)$	$S^2 = T^2 = (ST)^n = E$	§§ 1.5, 4.3
(l, m, n)	$R^l = S^m = T^n = RST = E$	§ 6.4
$\langle l, m, n\rangle$	$R^l = S^m = T^n = RST$	§ 6.5
$\langle l, m, n\rangle_q$	$R^l = S^m = T^n = RST = Z^q$, $RZ = ZR$, $SZ = ZS$, $TZ = ZT$	§ 6.5
$\langle l, m \mid n\rangle$	$R^l = S^m$, $(RS)^n = E$	§ 6.6
$\langle l, m \mid n; q\rangle$	$R^l = S^m = Z$, $(RS)^n = Z^q = E$	§ 6.6
$l\,[q]\,m$	$R^l = S^m = E$, $RSR\ldots = SRS\ldots$	§ 6.7
$(l, m \mid n, k)$	$R^l = S^m = (RS)^n = (R^{-1}S)^k = E$	§ 8.5
$(l, m, n; q)$	$R^l = S^m = T^n = RST = (TSR)^q = E$	§ 7.5
$G^{p,q,r}$	$A^p = B^q = C^r = (AB)^2 = (BC)^2 = (CA)^2 = (ABC)^2 = E$	§ 7.5
$[p, q]$	$R_1^2 = R_2^2 = R_3^2 = (R_1R_2)^p = (R_1R_3)^2 = (R_2R_3)^q = E$	§ 4.3
$[p, q]^+ \cong (2, p, q)$	$S^p = T^q = (ST)^2 = E$	§ 4.4
$[p^+, 2q]$	$S^p = T^2 = (S^{-1}TST)^q = E$	§ 4.4
$[p, q, r]$	$R_i^2 = (R_1R_2)^p = (R_2R_3)^q = (R_3R_4)^r = (R_1R_3)^2 = (R_1R_4)^2 = (R_2R_4)^2 = E$	§ 9.3
$[p, q, r]^+$	$T_1^p = T_2^q = T_3^r = (T_1T_2)^2 = (T_1T_2T_3)^2 = (T_2T_3)^2 = E$	§ 9.5
$[p^+, 2q, r^+]$	$S^p = T^r = (S^{-1}T^{-1}ST)^q = E$	§ 9.5

Appendix for Chapter 2

Example 1.

$\mathfrak{G} = \{B, C\},\ (BC)^2 = (B^3C^2)^2 = (B^2C^3)^2 = (B^{-1}C^2)^2 = E\,;$

$\mathfrak{H} = \{X, Y\},\ X = BC,\ Y = CB\,.$

The tables are set up, and with $1X = 1$, $1Y = 1$ inserted we have

	B	C
1		1

	C	B
1		1

	B	B	B	C	C	B	B	B	C	C
1										1

	B	C	B	C
1				1

Multiplication Table

	B	B^{-1}	C	C^{-1}
1				

	B	B	C	C	C	B	B	C	C	C
1										1

	B^{-1}	C	C	B^{-1}	C	C
1						1

New cosets will be defined by inserting the integer into the first available place in the multiplication table. Thus, we define $2 = 1B$, insert this into the multiplication table, start a new row indexed by 2 in each of the tables and transfer into them information from the multiplication table. We are immediately rewarded with $2C = 1$. We "process" this, define $3 = 1B^{-1}$, are rewarded with $1C = 3$ and the tables are:

	B	C
1	2	1

	C	B
1	3	1

	B	B	B	C	C	B	B	B	C	C
1	2								2	1
2										2
3	1	2				3	1	2	1	3

	B	C	B	C
1	2	1	2	1
2				2
3	1	3	1	3

	B	B^{-1}	C	C^{-1}
1	2	3	3	2
2		1	1	
3	1			1

	B	B	C	C	C	B	B	C	C	C
1	2								2	1
2										2
3	1	2	1	3				2	1	3

	B^{-1}	C	C	B^{-1}	C	C
1	3				2	1
2	1	3				2
3				2	1	3

Define, in turn, $4 = 2B$, $5 = 2C^{-1}$, $6 = 3B^{-1}$, $7 = 3C$, $8 = 4B$, and the tables are:

	B	C
1	2	1

	C	B
1	3	1

	B	B	B	C	C	B	B	B	C	C
1	2	4	8					5	2	1
2	4	8							5	2
3	1	2	4		6	3	1	2	1	3
4	8									4
5										5
6	3	1	2	1	3	1	2	4		6
7						6	3	1	3	7
8										8

	B	C	B	C
1	2	1	2	1
2	4		5	2
3	1	3	1	3
4	8			4
5				5
6	3	7		6
7		6	3	7
8				8

	B	B^{-1}	C	C^{-1}
1	2	3	3	2
2	4	1	1	5
3	1	6	7	1
4	8	2		
5			2	
6	3			
7				3
8		4		

	B	B	C	C	C	B	B	C	C	C
1	2	4								1
2	4	8								2
3	1	2	1	3	7					3
4	8									4
5										5
6	3	1	3	7						6
7		5	2	1	3	1	2	1	3	7
8										8

	B^{-1}	C	C	B^{-1}	C	C
1	3	7		5	2	1
2	1	3	7		5	2
3	6		4	2	1	3
4	2	1	3	6		4
5						5
6						6
7		5	2	1	3	7
8	4					8

Now define $9 = 4C$ and we are rewarded with $9C = 6$ (from row 3 of the $(B^3C^2)^2$ table), $9B = 5$ (from row 2 of the $(BC)^2$ table) and then $7B = 9$ (from row 6 of the $(BC)^2$ table). The computation proceeds smoothly with cosets defined by $10 = 4C^{-1}$ (the reward is $6C = 10$), $11 = 5B$ (the reward is $7C = 11$), $12 = 5C^{-1}$ (the reward is $7B^{-1} = 12$), $13 = 6B^{-1}$, $14 = 8B$, $15 = 8C$ (the rewards are $15C = 12$, $11B = 15$, $15B = 10$, $14C = 8$ and $13C = 14$) and $16 = 10B$ (the rewards are $16B = 12$, $16 = 11C$, $16C = 13$ and $14B = 13$). The tables now completely fill up. The information in the multiplication table is neatly expressed by the permutations

$$B\colon\ (1\ 2\ 4\ 8\ 14\ 13\ 6\ 3)\ (5\ 11\ 15\ 10\ 16\ 12\ 7\ 9)\,,$$

$$C\colon\ (1\ 3\ 7\ 11\ 16\ 13\ 14\ 8\ 15\ 12\ 5\ 2\ )\ (4\ 9\ 6\ 10)\,,$$

which show how the cosets are permuted when multiplied on the right by the generators of $\mathfrak{G}$.

With a little ingenuity we can see that the defining relations for $\mathfrak{G}$ imply $(B^2C^2)^3 = E$, whence $(XY)^3 = E$ so that $\mathfrak{H}$ is $\mathfrak{D}_3$ of order 6, and the order of $\mathfrak{G}$ is 96.

Example 2.

The alternating group $\mathfrak{A}_5$, of order 60, is generated by the permutations

$$S = (b\ e\ d)\,, \qquad U = (a\ b\ c\ d\ e)\,.$$

Since $SU = (a\ b)\ (c\,d)$, these generators satisfy

$$S^3 = U^5 = (SU)^2 = E\,,$$

defining an abstract group $\mathfrak{G}$ which we wish to show isomorphic to $\mathfrak{A}_5$. It suffices to show that the order of $\mathfrak{G}$ is at most 60.

Let $\mathfrak{H} = \{U\}$, a subgroup of $\mathfrak{G}$, whose order is at most 5 because $U^5 = 1$. Enumerate the cosets of $\mathfrak{H}$ in $\mathfrak{G}$. The tables with $1U = 1$ inserted are

	U	
1		1

	S		S		S	
1						1

	U		U		U		U		U	
1		1		1		1		1		1

	S		U		S		U	
1						1		1

	S	S^{-1}	U	U^{-1}
1			1	1

Define $2 = 1S$, $3 = 2S$; the S^3 table yields $3S = 1$, and the $(SU)^2$ table then yields $2U = 3$. The computation now stands as follows:

	U	
1		1

	S		S		S	
1		2		3		1

	U		U		U		U		U	
1		1		1		1		1		1
2		3								2

	S		U		S		U	
1		2		3		1		1
2		3						2

	S	S^{-1}	U	U^{-1}
1	2	3	1	1
2	3	1	3	
3	1	2		2

Continuing in this manner, we define, in order,

$$4 = 3U\,,\quad 5 = 4S\,,\quad 6 = 5S\,,\quad 7 = 4U\,,\quad 8 = 7S\,,\quad 9 = 8S\,,$$

$$10 = 9U\,,\quad 11 = 10S\,,\quad 12 = 11S\,.$$

When coset 12 has been defined the tables become

	S	S	S
1	2	3	1
4	5	6	4
7	8	9	7
10	11	12	10

	U	U	U	U	U
1	1	1	1	1	1
2	3	4	7	5	2
6	9	10	11	8	6
12	12	12	12	12	12

	S	U	S	U
1	2	3	1	1
2	3	4	5	2
6	4	7	8	6
9	7	5	6	9
10	11	8	9	10
12	10	11	12	12

The tables have "closed up" and hence the order of $\mathfrak{G}$ is at most $5 \times 12 = 60$.

Example 3.

$$\mathfrak{G} = \{A, B\}, \quad B^{-1}A^2B = A^3;$$

$$\mathfrak{H} = \{X, Y\}, \quad X = A^2, \; Y = B.$$

Set up the tables; of course $1 = \mathfrak{H}$ and $\bar{1} = E$.

	X^{-1}	A	A
1			1

	Y^{-1}	B
1		1

$$\bar{1}X = X\bar{1} \qquad \bar{1}Y = Y\bar{1}$$
$$\bar{1}X^{-1} = X^{-1}\bar{1} \qquad \bar{1}Y^{-1} = Y^{-1}\bar{1}$$

	B^{-1}	A	A	B	A^{-1}	A^{-1}	A^{-1}
1							1

	A	A^{-1}	B	B^{-1}
$\bar{1}$				

Insert the initial information and the row of the $Y^{-1}B$ table closes rewarding us with $\bar{1}B = \bar{1}Y = Y\bar{1}$. Store this in the information table, enter into the tables and we have

	X^{-1}	A	A
1	1		1

	Y^{-1}	B
1	1	1

$$\bar{1}X = X\bar{1} \qquad \bar{1}Y = Y\bar{1}$$
$$\bar{1}X^{-1} = X^{-1}\bar{1} \qquad \bar{1}Y^{-1} = Y^{-1}\bar{1}$$

	B^{-1}	A	A	B	A^{-1}	A^{-1}	A^{-1}
1	1						1

	A	A^{-1}	B	B^{-1}
$\bar{1}$			$Y\bar{1}$	$Y^{-1}\bar{1}$

Define $2 = 1A$, $\bar{2} = \bar{1}A = A$, store in the multiplication table (see below), start a new row indexed by 2 in the $B^{-1}A^2BA^{-3}$ table and enter

	A	A^{-1}	B	B^1
$\bar{1}$	$E\bar{2}$		$Y\bar{1}$	$Y^{-1}\bar{1}$
$\bar{2}$		$E\bar{1}$		

the information from the multiplication table. The row of the $X^{-1}A^2$ table closes and we find that

$$\bar{2}A = \bar{2}A^{-1}X = E\bar{1}X = X\bar{1}.$$

The multiplication table becomes

	A	A^{-1}	B	B^{-1}
$\bar{1}$	$E\bar{2}$	$X^{-1}\bar{2}$	$Y\bar{1}$	$Y^{-1}\bar{1}$
$\bar{2}$	$X\bar{1}$	$E\bar{1}$		

Now the first row of the $B^{-1}A^2BA^{-3}$ table closes:

	B^{-1}	A	A	B	A^{-1}	A^{-1}	A^{-1}
1	1	2	1	1	2	1	1
2							2

and we compute

$$\begin{aligned} \bar{1}A^{-1} &= \bar{1}AAB^{-1}A^{-1}A^{-1}B \\ &= E\bar{2}AB^{-1}A^{-1}A^{-1}B \\ &= X\bar{1}B^{-1}A^{-1}A^{-1}B \\ &= XY^{-1}\bar{1}A^{-1}A^{-1}B \\ &= XY^{-1}X^{-1}\bar{2}A^{-1}B \\ &= XY^{-1}X^{-1}E\bar{1}B \\ &= XY^{-1}X^{-1}Y\bar{1}\,. \end{aligned}$$

When we attempt to store this in $b(\bar{1}, A^{-1})$ we find $X^{-1}\bar{2}$ there and hence

$$XY^{-1}X^{-1}Y\bar{1} = X^{-1}\bar{2}$$

i. e.,

$$X^2Y^{-1}X^{-1}Y\bar{1} = \bar{2}\,, \quad Y^{-1}XYX^{-2}\bar{2} = \bar{1}\,.$$

Therefore we now replace 2 by 1 and $\bar{2}$ by $X^2Y^{-1}X^{-1}Y\bar{1}$ and the multiplication table becomes

	A	A^{-1}	B	B^{-1}
$\bar{1}$	$X^2Y^{-1}X^{-1}Y\bar{1}$	$XY^{-1}X^{-1}Y\bar{1}$	$Y\bar{1}$	$Y^{-1}\bar{1}$
$\bar{1}$	$Y^{-1}XYX^{-1}\bar{1}$	$Y^{-1}XYX^{-2}\bar{1}$		

Comparing corresponding entries in the two rows now indexed by $\bar{1}$ yields

$$X^2Y^{-1}X^{-1}Y\bar{1} = Y^{-1}XYX^{-1}\bar{1},$$

so

$$X^2Y^{-1}X^{-1}Y = Y^{-1}XYX^{-1}. \tag{A3.1}$$

The incomplete row of the $B^{-1}A^2BA^{-3}$ table, which was indexed by 2, is now indexed by 1 and it is filled in:

	B^{-1}	A	A	B	A^{-1}	A^{-1}	A^{-1}
1	1	1	1	1	1	1	1

This "closure" (underlined) yields

$$\begin{aligned} \bar{1}A^{-1} &= \bar{1}AAB^{-1}A^{-1}A^{-1}B \\ &= X^2Y^{-1}X^{-1}Y\bar{1}AB^{-1}A^{-1}A^{-1}B \\ &= (X^2Y^{-1}X^{-1}Y)^2\bar{1}B^{-1}A^{-1}A^{-1}B \\ &= (X^2Y^{-1}X^{-1}Y)^2Y^{-1}\bar{1}A^{-1}A^{-1}B \\ &= (X^2Y^{-1}X^{-1}Y)^2Y^{-1}(XY^{-1}X^{-1}Y)\bar{1}A^{-1}B \\ &= (X^2Y^{-1}X^{-1}Y)^2Y^{-1}(XY^{-1}X^{-1}Y)^2\bar{1}B \\ &= (X^2Y^{-1}X^{-1}Y)^2Y^{-1}(XY^{-1}X^{-1}Y)^2Y\bar{1}\,. \end{aligned}$$

When we try to put this into $b(\bar{1}, A^{-1})$ we find $XY^{-1}X^{-1}Y\bar{1}$ there, and hence

$$(X^2Y^{-1}X^{-1}Y)^2Y^{-1}(XY^{-1}X^{-1}Y)^2Y = XY^{-1}X^{-1}Y. \qquad \text{(A3.2)}$$

Relations A 3.1 and A 3.2 suffice to define $\mathfrak{H} = \mathfrak{G}$. Since the former implies the latter, $\mathfrak{G}$ is defined by the single relation

$$X = (Y^{-1}XYX^{-1})^2.$$

Note that this relation implies $Y^{-1}X^2Y = X^3$ but not conversely. The reader may wish to show that the relations

$$B^{-1}A^2B = A^3, \quad X = A^2, \quad Y = B$$

are equivalent to

$$X = (Y^{-1}XYX^{-1})^2, \quad A = Y^{-1}XYX^{-1}, \quad B = Y.$$

Example 4.

$$\mathfrak{G} = \{A, B\}, \quad A^2BAB^3 = B^2ABA^3 = E;$$
$$\mathfrak{H} = \{X\}, \qquad X = A.$$

Set up the tables with $1 = \mathfrak{H}$ and $\bar{1} = E$.

	X^{-1}	A
1		1

Table I

	A	A	B	A	B	B	B
1							1

Table II

	B	B	A	B	A	A	A
1							1

$\bar{1}X = X\bar{1}$, $\bar{1}X^{-1} = X^{-1}\bar{1}$

	A	A^{-1}	B	B^{-1}
$\bar{1}$				

Inserting $1X^{-1} = 1$ yields $\bar{1}A = X\bar{1}$. Cosets and representatives are then defined as follows: $2 = 1B$, $\bar{2} = B$; $3 = 1B^{-1}$, $\bar{3} = B^{-1}$; $4 = 2B$, $\bar{4} = B^2$. At this point row 1 of Table II closes to yield

$$\bar{4}A = \bar{4}B^{-1}B^{-1}A^{-1}A^{-1}A^{-1}B^{-1} = \bar{2}B^{-1}A^{-1}A^{-1}B^{-1} = \ldots = X^{-3}\bar{3},$$

and then row 4 of Table I closes, showing that

$$\begin{aligned}\bar{3}A &= \bar{3}A^{-1}B^{-1}B^{-1}B^{-1}A^{-1}B^{-1}\\ &= X^3\bar{4}B^{-1}B^{-1}B^{-1}A^{-1}B^{-1}\\ &= X^3\bar{3}A^{-1}B^{-1} = X^3X^3\bar{4}B^{-1} = X^6\,\bar{2}.\end{aligned}$$

Define $5 = 2A$, $\bar{5} = BA$ and at this stage the computation stands at

	X^{-1}	A
1	1	1

Table I

	A	A	B	A	B	B	B
1	1	1	2	5		3	1
2	5				3	1	2
3	2	5					3
4	3	2	4	3	1	2	4
5							5

Table II

	B	B	A	B	A	A	A
1	2	4	3	1	1	1	1
2	4				4	3	2
3	1	2	5			4	3
4							4
5		3	2	4	3	2	5

	A	A^{-1}	B	B^{-1}
$\bar{1}$	$X\bar{1}$	$X^{-1}\bar{1}$	$E\bar{2}$	$E\bar{3}$
$\bar{2}$	$E\bar{5}$	$X^{-6}\bar{3}$	$E\bar{4}$	$E\bar{1}$
$\bar{3}$	$X^{6}\bar{2}$	$X^{3}\bar{4}$	$E\bar{1}$	
$\bar{4}$	$X^{-3}\bar{3}$			$E\bar{2}$
$\bar{5}$		$E\bar{2}$		

Now define $6 = 3B^{-1}$, $\bar{6} = B^{-2}$ and we see that row 1 of Table I closes to yield $\bar{5}B = X^{-2}\bar{6}$ (enter this into the multiplication table) while row 5 of Table II closes to yield $\bar{5}B = X^{-9}\bar{6}$. Hence $X^{-2}\bar{6} = X^{-9}\bar{6}$ i. e.,

$$X^{7} = E\,.$$

We continue by defining $7 = 4A^{-1}$, $\bar{7} = B^{2}A^{-1}$; from row 3 of II we deduce $\bar{6}\,A = X^{5}\bar{7}$ and then row 3 of I yields $\bar{7}B = X^{-7}\bar{5}$ followed by $\bar{4}B = X^{2}\bar{7}$ from row 7 of I. Finally, taking $8 = 5A$, $\bar{8} = BA^{2}$ we deduce from row 5 of I that $\bar{8}A = X^{-1}\bar{6}$, and from row 2 of I that $\bar{8}B = X\bar{8}$. All incomplete rows are now filled in; the resulting relations are all deducible from $X^{7} = E$. Hence $\mathfrak{H}$ is presented by $X^{7} = E$ and the index of $\mathfrak{H}$ in $\mathfrak{G}$ is 8.

The complete multiplication table is

	A	A^{-1}	B	B^{-1}
$\bar{1}$	$X\bar{1}$	$X^{-1}\bar{1}$	$E\bar{2}$	$E\bar{3}$
$\bar{2}$	$E\bar{5}$	$X^{-6}\bar{3}$	$E\bar{4}$	$E\bar{1}$
$\bar{3}$	$X^{6}\bar{2}$	$X^{3}\bar{4}$	$E\bar{1}$	$E\bar{6}$
$\bar{4}$	$X^{-3}\bar{3}$	$E\bar{7}$	$X^{2}\bar{7}$	$E\bar{2}$
$\bar{5}$	$E\bar{8}$	$E\bar{2}$	$X^{-2}\bar{6}$	$X^{7}\bar{7}$
$\bar{6}$	$X^{5}\bar{7}$	$X\bar{8}$	$E\bar{3}$	$X^{2}\bar{5}$
$\bar{7}$	$E\bar{4}$	$X^{-5}\bar{6}$	$X^{-7}\bar{5}$	$X^{-2}\bar{4}$
$\bar{8}$	$X^{-1}\bar{6}$	$E\bar{5}$	$X\bar{8}$	$X^{-1}\bar{8}$

Bibliography

ARTIN, E.:

1926 Theorie der Zöpfe, Abh. Math. Sem. Univ. Hamburg **4**, 47–72.

1947a Theory of braids, Ann. of Math. (2), **48**, 101–126.

1947b Braids and permutations, Ann. of Math. (2), **48**, 643–649.

1955 The orders of the classical simple groups, Comm. Pure Appl. Math. (New York) **8**, 455–472.

BAER, R.:

1944 The higher commutator subgroups of a group, Bull. Amer. Math. Soc. **50**, 143–160.

1945 Representation of groups as quotient groups I, Trans. Amer. Math. Soc. **58**, 295–347.

— and F. LEVI:

1936 Freie Produkte und ihre Untergruppen, Comp. Math. **3**, 391–398.

BAKER, H. F.:

1946 A locus with 25920 linear self-transformations, Cambridge.

BAKER, R. P.:

1931 Cayley diagrams on the anchor ring, Amer. J. Math. **53**, 645–669.

BALL, W. W. R.:

1974 Mathematical Recreations and Essays (12th ed.), Toronto.

BAMBAH, R. P., and H. DAVENPORT:

1952 The covering of n-dimensional space by spheres, J. London Math. Soc. **27**, 224–229.

BEETHAM, M. J., and C. M. CAMPBELL:

1976 A note on the Todd-Coxeter coset enumeration algorithm, Proc. Edinburgh Math. Soc. (2) **20**, 73–79.

BEHR, H., and J. MENNICKE:

1968 A presentation of the groups $PSL\ (2, p)$, Canad. J. Math. **20**, 1432–1438.

BELOWA, E. N., N. W. BELOW and A. SCHUBNIKOW:

1948 On the number and character of the abstract groups corresponding to the 32 crystallographic classes, Doklady Akad. Nauk SSSR **63**, 669–672.

BENSON, C. T., and N. S. MENDELSOHN:

1966 A calculus for a certain class of word problems in groups, J. Comb. Theory **1**, 202–208.

BILINSKI, S.:

1950 Homogene mreže zatvorenih orijentabilnih ploha. Rad Jugoslav. Akad. Znan. Umjet. Odjel Mat. Fiz. Tehn. Nauke **277**, 129–164.

1952 Homogene Netze geschlossener orientierbarer Flächen. Bull. Internat. Acad. Yougoslave, Cl. Sci. Math. Phys. Tech. (N. S.) **6**, 59–75.

BLICHFELDT, H. F.:

1929 The minimum value of quadratic forms and the closest packing of spheres, Math. Ann. **101**, 605–608.

BOHNENBLUST, F.:

1947 The algebraical braid group, Ann. of Math. (2), **48**, 127–136.

BRAHANA, H. R.:

1926 Regular maps on an anchor ring, Amer. J. Math. **48**, 225–240.

1927 Regular maps and their groups, Amer. J. Math. **49**, 268–284.

— and A. B. COBLE:

1926 Maps of twelve countries with five sides with a group of order 120 containing an icosahedral subgroup, Amer. J. Math. **48**, 1–20.

BUNDGAARD, S.:
1952 On a kind of homotopy in regular numbered complexes, Medd. Lunds Univ. Mat. Sem., Tome Suppl. 35–46.

BURCKHARDT, J. J.:
1947 Die Bewegungsgruppen der Kristallographie, Basel.

BURNS, J. E.:
1915 Abstract definitions of groups of degree eight, Amer. J. Math. **37**, 195–214.

BURNSIDE, W.:
1897 Note on the symmetric group, Proc. London Math. Soc. (1), **28**, 119–129.
1899 Note on the simple group of order 504, Math. Ann. **52**, 174–176.
1902 On an unsettled problem in the theory of discontinuous groups, Quart. J. Math. **33**, 230–238.
1911 Theory of Groups of Finite Order (2nd ed.), Cambridge.
1912 The determination of all groups of rational linear substitutions of finite order which contain the symmetric group in the variables, Proc. London Math. Soc. (2), **10**, 284–308.

BUSSEY, W. H.:
1905 Generational relations for the abstract group simply isomorphic with the group $LF[2, p^n]$, Proc. London Math. Soc. (2), **3**, 296–315.

CAMPBELL, C. M.:
1970 Some examples using coset enumeration, Computational Problems in Abstract Algebra, Proceedings of a Conference, Oxford, 1967.

CANNON, J. J., L. A. DIMINO, G. HAVAS, and J. M. WATSON:
1973 Implementation and Analysis of the Todd-Coxeter Algorithm, Math. of Computation 27, 463—490.

CARMICHAEL, R. D.:
1923 Abstract definitions of the symmetric and alternating groups and certain other permutation groups, Quart. J. Math. **49**, 226–270.
1937 Introduction to the Theory of Groups of Finite Order, Boston.

CARTAN, E.:
1927 La géométrie des groupes simples, Ann. Mat. Pura Appl. (4), 209–256.
1928 Complément au mémoire sur la géométrie des groupes simples, Ann. Mat. Pura Appl. (4), **5**, 253–260.

CAYLEY, A.:
1878a On the theory of groups, Proc. London Math. Soc. (1), **9**, 126 to 133.
1878b The theory of groups: graphical representations, Amer. J. Math. **1**, 174–176.
1889 On the theory of groups, Amer. J. Math. **11**, 139–157.

CHEN, K. T.:
1951 Integration in free groups, Ann. of Math. **54**, 147–162.
1954 A group ring method for finitely generated groups, Trans. Amer. Math. Soc. **76**, 275–287.

CHEVALLEY, C.:
1955a Invariants of finite groups generated by reflections, Amer. J. Math. **77**, 778–782.
1955b Sur certains groupes simples, Tôhoku Math. J. (2), **7**, 14–66.

CHOW, W.:
1948 On the algebraical braid group, Ann. of Math. (2), **49**, 654–658.

COLEMAN, A. J.:
1958 The Betti numbers of simple Lie groups, Canad. J. Math. **10**, 349—356.
CONWAY, J. H.:
1971 Three lectures on exceptional groups. Finite Simple Groups, 215—247, (Proceedings of an Instructional Conference, Oxford, 1969), London.
COX, H.:
1891 Application of Grassmann's Ausdehnungslehre to properties of circles, Quart. J. Math. **25**, 1—71.
COXETER, H. S. M.:
1928 The pure Archimedean polytopes in six and seven dimensions, Proc. Cambridge Philos. Soc. **24**, 1—9.
1931 Groups whose fundamental regions are simplexes, J. London Math. Soc. **6**, 132—136.
1934a Discrete groups generated by reflections, Ann. of Math. **35**, 588—621.
1934b On simple isomorphism between abstract groups, J. London Math. Soc. **9**, 211—212.
1934c Abstract groups of the form $V_1^k = V_j^3 = (V_i V_j)^2 = 1$, J. London Math. Soc. **9**, 213—219.
1935 The complete enumeration of finite groups of the form $R_i^2 = (R_i R_j)^{k_{ij}} = 1$, J. London Math. Soc. **10**, 21—25.
1936a The groups determined by the relations $S^l = T^m = (S^{-1}T^{-1}ST)^p = 1$, Duke Math. J. **2**, 61—73.
1936b The abstract groups $R^m = S^m = (R^j S^j)^{p_j} = 1$, $S^m = T^2 = (S^j T)^{2p_j} = 1$, $S^m = T^2 = (S^{-j}TS^jT)^{p_j} = 1$, Proc. London Math. Soc. (2), **41**, 278—301.
1936c An abstract definition for the alternating group in terms of two generators, J. London Math. Soc. **11**, 150—156.
1937a Regular skew polyhedra in three and four dimensions and their topological analogues, Proc. London Math. Soc. (2), **43**, 33—62.
1937b Abstract definitions for the symmetry groups of the regular polytopes in terms of two generators. Part II: The rotation groups, Proc. Cambridge Philos. Soc. **33**, 315—324.
1939 The abstract groups $G^{m,n,p}$, Trans. Amer. Math. Soc. **45**, 73—150.
1940a Regular and semi-regular polytopes I, Math. Z. **46**, 380—407.
1940b The polytope 2_{21}, whose twenty-seven vertices correspond to the lines on the general cubic surface, Amer. J. Math. **62**, 457—486.
1940c The binary polyhedral groups and other generalizations of the quaternion group, Duke Math. J. **7**, 367—379.
1940d A method for proving certain abstract groups to be infinite, Bull. Amer. Math. Soc. **46**, 246—251.
1946a Quaternions and reflections, Amer. Math. Monthly **53**, 136—146.
1946b Integral Cayley numbers, Duke Math. J. **13**, 561—578.
1948 Configurations and maps, Rep. Math. Colloq. (2), **8**, 18—38.
1950 Self-dual configurations and regular graphs, Bull. Amer. Math. Soc. **56**, 413—455.
1951a Extreme forms, Canad. J. Math. **3**, 391—441.
1951b The product of generators of a finite group generated by reflections, Duke Math. J. **18**, 765—782.

COXETER, H. S. M.:

1954 Regular honeycombs in elliptic space, Proc. London Math. Soc. (3), **4**, 471–501.

1955 Affine geometry, Scripta Math. **21**, 5–14.

1956 The collineation groups of the finite affine and projective planes with four lines through each point, Abh. Math. Sem. Univ. Hamburg, **20**, 165–177.

1957a Non-Euclidean Geometry (3rd ed.), Toronto.

1957b Groups generated by unitary reflections of period two, Canad. J. Math. **9**, 243–272.

1958a Twelve points in $PG(5, 3)$ with 95040 self-transformations, Proc. Roy. Soc. London, A 247, 279–293.

1958b On the subgroups of the modular group, J. de Math. pures appl. **37**, 317–319.

1959a Factor groups of the braid group, Proc. Fourth Canad. Math. Congress, 95–122.

1959b Symmetrical definitions for the binary polyhedral groups, Proceedings of Symposia in Pure Mathematics (American Mathematical Society) **1**, 64–87.

1961 Introduction to Geometry, New York, (2nd ed., 1969.)

1962a The abstract group $G^{3,7,16}$, Proc. Edinburgh Math. Soc. **13** (II), 47–61 and 189.

1962b The symmetry groups of the regular complex polygons, Arch. Math. **13**, 86–97.

1962c The classification of zonohedra by means of projective diagrams, J. de Math. pures appl. **41**, 137–156.

1963a Regular Polytopes, New York, (3rd. ed., 1973.)

1963b Unvergängliche Geometrie, Basel (Translation of COXETER 1961).

1970 Twisted honeycombs. Regional Conference Series in Mathematics, No. 4, American Mathematical Society, Providence, R. I.

1974 Regular Complex Polytopes, Cambridge.

— M. S. LONGUET-HIGGINS and J. C. P. MILLER:

1954 Uniform polyhedra, Philos. Trans. Roy. Soc. London, A **246**, 401–450.

— and J. A. TODD:

1936 Abstract definitions for the symmetry groups of the regular polytopes in terms of two generators. Part I: The complete groups, Proc. Cambridge Philos. Soc. **32**, 194–200.

— and G. J. WHITROW:

1950 World Structure and non-Euclidean honeycombs, Proc. Roy. Soc. London, A **201**, 417–437.

DEDEKIND, R.:

1897 Über Gruppen, deren sämtliche Teiler Normalteiler sind, Math. Ann. **48**, 548–561.

DEHN, M.:

1910 Über die Topologie des dreidimensionalen Raumes, Math. Ann. **69**, 137–168.

1912 Über unendliche diskontinuierliche Gruppen, Math. Ann. **71**, 116–144.

DE SÉGUIER, *see* SÉGUIER.

DICKSON, L. E.:
1901a Linear Groups, with an Exposition of the Galois Field theory, Leipzig.
1901b Theory of linear groups in an arbitrary field, Trans. Amer. Math. Soc. **2**, 363–394.
1903 The abstract group G simply isomorphic with the alternating group on six letters, Bull. Amer. Math. Soc. **9**, 303–306.
1905 A new system of simple groups, Math. Ann. **60**, 137–150.
DIEUDONNÉ, J.:
1954 Les isomorphismes exceptionnels entre les groupes classiques finis, Canad. J. Math. **6**, 305–315.
1955 La Géométrie des Groupes Classiques, Erg. d. Math. N. F. **5**.
DOLIVO-DOBROVOLSKY, V. V.:
1925 Recherches sur le système dodécaèdre-icosaèdrique, Mem. Soc. Russe Min. **52**, 169–181.
DU VAL, P.:
1933 On the directrices of a set of points in a plane, Proc. London Math. Soc. (2), **35**, 23–74.
DYCK, W.:
1880 Über Aufstellung und Untersuchung von Gruppe und Irrationalität regularer Riemannscher Flächen, Math. Ann. **17**, 473–508.
1882 Gruppentheoretische Studien, Math. Ann. **20**, 1–45.
DYE, D. S.:
1937 A grammar of Chinese lattice, Cambridge, Mass.
EDGE, W. L.:
1955 The isomorphism between $LF(2, 3^2)$ and $\mathfrak{A}_6$, J. London Math. Soc. **30**, 172–185.
1963 An orthogonal group of order $2^{13} \cdot 3^5 \cdot 5^2 \cdot 7$, Annali di Mat. (4), **61**, 1–95.
ERRÉRA, A.:
1922 Sur les polyèdres réguliers de l'Analysis Situs, Acad. Roy. Belg. Cl. Sci. Mem. Coll. in 8° (2), **7**, 1–17.
ESCHER, M. C.:
1961 The Graphic Work of M. C. Escher, London.
FEDOROV, E. S.:
1885 1953. Načala Učenija o Figurach, Leningrad.
1891 Zapiski Mineralogicheskogo Imperatorskogo S. Petersburgskogo Obshchestva (2), **28**, 345–390.
FEJES TÓTH, L.:
1953 Lagerungen in der Ebene, auf der Kugel und im Raum, Grundlehren der mathematischen Wissenschaften **65**, Berlin.
FELSCH, H:
1961 Programmierung der Restklassenabzählung einer Gruppe nach Untergruppen, Numer. Math. **3**, 250—256.
FORD, L. R.:
1929 Automorphic functions, New York.
FOX, R. H.:
1953 Free differential calculus I, Ann. of Math. (2), **57**, 547–560.
1954 Free differential calculus II, Ann. of Math. (2), **59**, 196–210.
FRASCH, H.:
1933 Die Erzeugenden der Hauptkongruenzgruppen für Primzahlstufen, Math. Ann. **108**, 229–252.

FRICKE, R.:
1892 Über den arithmetischen Charakter der zu den Verzweigungen (2, 3, 7) und (2, 4, 7) gehörenden Dreiecksfunktionen, Math. Ann. **41**, 443—468.

— and F. KLEIN:
1897 Vorlesungen über die Theorie der automorphen Funktionen, Leipzig.

FRUCHT, R.:
1955 Remarks on finite groups defined by generating relations, Canad. J. Math. **7**, 8—17, 413.

FRYER, K. D.:
1955 A class of permutation groups of prime degree, Canad. J. Math. **7**, 24—34.

GOURSAT, E.:
1889 Sur les substitutions orthogonales et les divisions régulières de l'espace, Ann. Sci. Ecole Norm. Sup. (3), **6**, 9—102.

GREEN, J. A.:
1952 On groups with odd prime-power exponent, J. London Math. Soc. **27**, 476—485.

GRÜN, O.:
1936 Über eine Faktorgruppe freier Gruppen. I, Dtsch. Math. **1**, 772—782.
1940 Zusammenhang zwischen Potenzbildung und Kommutatorbildung, J. reine angew. Math. **182**, 158—177.

HALL, M.:
1949 Coset representations in free groups, Trans. Amer. Math. Soc. **67**, 421—432.
1958 Solution of the Burnside problem for exponent six, Illinois J. Math. **2**, 764—786.
1959 The Theory of Groups, New York.
1962 Note on the Mathieu group M_{12}, Arch. Math. **13**, 334—340.

— and J. K. SENIOR:
1964 The Groups of Order 2^n, $n \leqq 6$, New York.

HALL, P.:
1933 A contribution to the theory of groups of prime-power order, Proc. London Math. Soc. (2), **36**, 29—95.

— and G. HIGMAN:
1956 On the p-length of p-soluble groups and reduction theorems for Burnside's problem, Proc. London Math. Soc. (3), **6**, 1—42.

HAMILL, C. M.:
1951 On a finite group of order 6,531,840, Proc. London Math. Soc. (2), **52**, 401—454.

HAMILTON, W. R.:
1856 Memorandum respecting a new system of roots of unity, Philos. Mag. (4), **12**, 446.

HAVAS, G.:
1974a A Reidemeister-Schreier program, Proc. Second Internat. Conf. Theory of Groups, Canberra 1973, 347—352. (Lecture Notes in Math., 372, Springer-Verlag, Berlin 1974).
1974b Computational approaches to combinatorial group theory. Ph. D. Thesis, University of Sydney. (Bull. Austral. Math. Soc. **11** (1974), 475—476).

HEAWOOD, P. J.:
1890 Map-colour theorem, Quart. J. Math. **24**, 332—338.
HEESCH, H., and O. KIENZLE:
1963 Flächenschluß. System der Formen lückenlos aneinanderschließender Flachteile, Berlin.
HEFFTER, L.:
1891 Über das Problem der Nachbargebiete, Math. Ann. **38**, 477—508.
1898 Über metacyklische Gruppen und Nachbarconfigurationen, Math. Ann. **50**, 261—268.
HENRY, N. F. M., and K. LONSDALE:
1952 International Tables for X-ray Crystallography I, Birmingham.
HERMANN, C.:
1949 Kristallographie in Räumen beliebiger Dimensionszahl I, Acta Crystallogr. **2**, 139—145.
HESS, E.:
1876 Über die zugleich gleicheckigen und gleichflächigen Polyeder, Schr. Ges. Beförd. Naturwiss. Marburg **11**, 1.
HESSEL, J. F. C.:
1897 Krystallometrie oder Krystallonomie und Krystallographie (Ostwald's Klassiker der exakten Wissenschaften **88**, **89**), Leipzig.
HIGMAN, G.:
1956 On finite groups of exponent five, Proc. Cambridge Philos. Soc. **52**, 381—390.
1957 Le problème de Burnside. Colloque d'Algèbre supérieure, tenu a Bruxelles du 19 au 22 décembre 1956, Centre Belge de Recherches Mathématique, Paris.
HILBERT, D., and S. COHN-VOSSEN:
1932 Anschauliche Geometrie, Berlin.
HILTON, H.:
1908 The Theory of Groups of Finite Order, Oxford.
HINTON, C. H.:
1906 The Fourth Dimension, London.
HUA, L. K., and I. REINER:
1949 On the generators of the symplectic group, Trans. Amer. Math. Soc. **65**, 415—426.
1951 Automorphisms of the unimodular group, Trans. Amer. Math. Soc. **71**, 331—348.
1952 Automorphisms of the projective unimodular group, Trans. Amer. Math. Soc. **72**, 467—473.
HUDSON, R. W. H. T.:
1905 Kummer's Quartic Surface, Cambridge.
HURLEY, A. C.:
1951 Finite rotation groups and crystal classes in four dimensions, Proc. Cambridge Philos. Soc. **47**, 650—661.
JACOBSON, N.:
1943 The Theory of Rings (Mathematical Surveys **2**), New York.
JOHNSON, D. L.:
1976 Presentation of Groups, Cambridge Univ. Press.
JORDAN, C.:
1870 Traité des Substitutions, Paris.

KELVIN, LORD:
1894 On homogeneous divisions of space, Proc. Roy. Soc. London, A **55**, 1–16.

KEMPE, A. B.:
1886 A memoir on the theory of mathematical form, Philos. Trans. Roy. Soc. London, A **177**, 1–70.

KEPLER, J.:
1619 Harmonice Mundi, Opera omnia **5** (Frankfurt, 1864).

KERÉKJÁRTÓ, B.:
1923 Vorlesungen über Topologie **1**, Berlin.

KILLING, W.:
1888 Die Zusammensetzung der stetigen endlichen Transformationsgruppen II, Math. Ann. **33**, 1–48.

KLEIN, F.:
1876 Über binäre Formen mit linearen Transformationen in sich selbst, Math. Ann. **9**, 183–208.
1879a Über die Transformation der elliptischen Functionen und die Auflösung der Gleichungen fünften Grades, Math. Ann. **14**, 111–172.
1879b Über die Transformation siebenter Ordnung der elliptischen Functionen, Math. Ann. **14**, 428–471.
1884 Vorlesungen über das Ikosaeder und die Auflösung der Gleichungen fünften Grades, Leipzig.

— and R. FRICKE:
1890 Vorlesungen über die Theorie der elliptischen Modulfunktionen, Leipzig.

KÖNIG, D.:
1936 Theorie der endlichen und unendlichen Graphen, Leipzig.

KORKINE, A., and G. ZOLOTAREFF:
1873 Sur les formes quadratiques, Math. Ann. **6**, 366–389.

KOSTANT, B.:
1959 The principal three-dimensional subgroup and the Betti numbers of a complex simple Lie group, Amer. J. Math. **81**, 973–1032.

KOSTRIKIN, A. I.:
1955 Solution of the restricted Burnside Problem for the exponent 5, Isvestiya Akad. Nauk SSSR **19**, 233–244.

KUROSCH, A. G.:
1953 Gruppentheorie, Berlin.

LANNÉR, F.:
1950 On complexes with transitive groups of automorphisms, Medd. Lunds Univ. Math. Sem. **11**, 1–71.

LEECH, J.:
1962 Some definitions of Klein's simple group of order 168 and other groups, Proc. Glasgow Math. Assoc. **5**, 166–175.
1963 Coset enumeration on digital computers, Proc. Camb. Phil. Soc. **59**, 257–267.

LEECH, J.:

1969 A presentation of the Mathieu group M_{12}, Canad. Math. Bull. **12**, 41—43.

1970 Computational Problems in Abstract Algebra, Proceedings of a Conference, Oxford, 1967.

1977 Computer proof of relations in groups. Topics in Group Theory and Computations (Proc. Royal Irish Acad. Summer School on Group Theory and Computation, Galway, 1973, Academic Press, 1977).

LEFSCHETZ, S.:

1949 Introduction to Topology, Princeton.

LEVI, F. W., and B. L. VAN DER WAERDEN:

1933 Uber eine besondere Klasse von Gruppen, Abh. Math. Sem. Univ. Hamburg **9**, 154—158.

LEWIS, F. A.:

1938 Note on the defining relations for the simple group of order 660, Bull. Amer. Math. Soc. **44**, 456.

LINDSAY, J. H.:

1959 An elementary treatment of the imbedding of a graph in a surface, Amer. Math. Monthly **66**, 117—118.

LYNDON, R. C.:

1950 Cohomology theory of groups with a single defining relation, Ann. of Math. **52**, 650—665.

1954 On Burnside's problem, Trans. Amer. Math. Soc. **77**, 202—215.

MACDUFFEE, C. C.:

1933 The Theory of Matrices, Erg. d. Math. **2**, 5.

MAGNUS, W.:

1932 Das Identitätsproblem für Gruppen mit einer definierenden Relation, Math. Ann. **106**, 295—307.

1935a Beziehungen zwischen Gruppen und Idealen in einem speziellen Ring, Math. Ann. **111**, 259—280.

1935b Uber n-dimensionale Gittertransformationen, Acta Mathematica **64**, 355—367.

1937 Uber Beziehungen zwischen höheren Kommutatoren, J. reine angew. Math. **177**, 105—115.

1939 Uber freie Faktorgruppen und freie Untergruppen gegebener Gruppen, Mh. Math. Phys. **47**, 307—313.

1950 A connection between the Baker-Hausdorff formula and a problem of Burnside, Ann. of Math. (2), **52**, 111—126.

— A. KARRASS and S. SOLITAR

1966 Combinatorial Group Theory: Presentation of Groups in Terms of Generators and Relations, New York.

MASCHKE, H.:

1896 The representation of finite groups, especially of the rotation groups of the regular bodies in three- and four-dimensional space, by Cayley's Color Diagrams, Amer. J. Math. **18**, 156—194.

MATHIEU, E.:

1861 Mémoire sur l'étude des fonctions de plusiers quantités, J. de Math. (2), **6**, 241—323.

1873 Sur la fonction cinq fois transitive de 24 quantités, J. de Math. (2), **18**, 25—46.

MEIER-WUNDERLI, H.:
1956 Uber die Struktur der Burnsidegruppen mit zwei Erzeugenden und vom Primzahlexponenten $p > 3$, Comment. Math. Helvet. **30**, 144–174.

MENDELSOHN, N. S.:
1970 Defining relations for subgroups of finite index of groups with a finite presentation, Computational Problems in Abstract Algebra (Proc. Conference, Oxford) Pergamon Press.

MENNICKE, J. L., and D. GARBE:
1964 Some remarks on the Mathieu groups, Canad. Math. Bull. **7**, 201—212.

MILLER, G. A.:
1901 On the groups generated by two operators of orders two and three respectively whose product is of order six, Quart. J. Math. **33**, 76–79.
1902 Groups defined by the orders of two generators and the order of their product, Amer. J. Math. **24**, 96–100.
1907 Generalization of the groups of genus zero, Trans. Amer. Math. Soc. **8**, 1–13.
1908 The groups generated by two operators which have a common square, Arch. Math. Phys. **9**, 6–7.
1909 Finite groups which may be defined by two operators satisfying two conditions, Amer. J. Math. **31**, 167–182.
1911 Abstract definitions of all the substitution groups whose degrees do not exceed seven, Amer. J. Math. **33**, 363–372.
1920 Groups generated by two operators of order three whose product is of order four, Bull. Amer. Math. Soc. **26**, 361–369.
1930 Determination of all the groups of order 64, Amer. J. Math. **52**, 617–634.

— H. F. BLICHFELDT and L. E. DICKSON:
1916 Theory and Application of Finite Groups, New York.

— and H. C. MORENO:
1903 Non-abelian groups in which every subgroup is abelian, Trans. Amer. Math. Soc. **4**, 398–404.

MITCHELL, H. H.:
1914 Determination of all primitive collineation groups in more than four variables which contain homologies, Amer. J. Math. **36**, 1–12.

MOORE, E. H.:
1897 Concerning the abstract groups of order $k!$ and $\frac{1}{2}k!$ . . ., Proc. London Math. Soc. (1), **28**, 357–366.

MOSER, W. O. J.:
1959 Abstract definitions for the Mathieu groups M_{11} and M_{12}, Canad. Math. Bull. **2**, 9–13.
1964 Remarks on a paper by Trott, Canad. Math. Bull. **7**, 49–52.

MÜLLER, E.:
1944 Gruppentheoretische und strukturanalytische Untersuchung der Maurischen Ornamente aus der Alhambra in Granada, Rüschlikon.

NETTO, E.:
1900 Vorlesungen über Algebra, Leipzig.

NEUMANN, B. H.:
1932 Die Automorphismengruppe der freien Gruppen, Math. Ann. **107**, 367–386.
1937a Groups whose elements have bounded orders, J. London Math. Soc. **12**, 195–198.
1937b Identical relations in groups I, Math. Ann. **114**, 506–525.
1956 On some finite groups with trivial multiplicator, Publ. Math. Debrecen **4**, 190–194.

— and HANNA NEUMANN:
1951 Zwei Klassen charakteristischer Untergruppen und ihre Faktorgruppen, Math. Nachr. **4**, 106–125.

NIELSEN, J.:
1921 Om regning med ikkekommutative faktorer og dens anvendelse i gruppeteorien, Mat. Tidsskr. B, 77–94.
1924a Die Gruppe der dreidimensionalen Gittertransformationen, Danske Vid. Selsk. Mat.-Fys. Medd. **5.12** (29 pp.).
1924b Die Isomorphismengruppen der freien Gruppen, Math. Ann. **91**, 169–209.
1927 Untersuchungen zur Topologie der geschlossenen zweiseitigen Flächen I, Acta math. **50**, 189–358.
1932 Untersuchungen zur Topologie der geschlossenen zweiseitigen Flächen III, Acta math. **58**, 87–167.
1940 Die symmetrische und die alternierende Gruppe, Mat. Tidsskr. B 7–18.
1950 A study concerning the congruence subgroups of the modular group, Danske Vid. Selsk. Mat.-Fys. Medd. **25.18** (32 pp.).

NIGGLI, P.:
1924 Die Flächensymmetrien homogener Diskontinuen, Z. Kristallogr., Mineralog. Petrogr. Abt. A **60**, 283–298.

NOVIKOV, P. S., and S. I. ADJAN:
1968 Infinite periodic groups I, II, III, Izv. Akad. Nauk SSSR Ser. Mat. **32**, 212—244, 251—524, 709—731 (= Math USSR Izv. **2**, 209—236, 241—479, 665—685).

NOWACKI, W.:
1933 Die nichtkristallographischen Punktgruppen, Z. Kristallogr., Mineralog. Petrogr. Abt. A **86**, 19–31.
1954 Über die Anzahl verschiedener Raumgruppen, Schweiz. Mineral. Petrogr. Mitt. **34**, 130–168.

O'CONNOR, R. E., and G. PALL:
1944 The construction of integral quadratic forms of determinant 1, Duke Math. J. **11**, 319–331.

ORE, O.:
1962 Theory of Graphs, American Mathematical Society Colloqium Publication XXXXIII.

PALEY, R. E. A. C.:
1933 On orthogonal matrices, J. Math. Phys. **12**, 311–320.

PASCAL, E.:
1927 Repertorium der höheren Mathematik I_2, Leipzig.

POINCARÉ, H.:
1882 Théorie des groupes fuchsiens, Acta Math. **1**, 1–62.

PÓLYA, G.:
1924 Über die Analogie der Kristallsymmetrie in der Ebene, Z. Kristallogr., Mineralog. Petrog. Abt. A **60**, 278–282.
1937 Kombinatorische Anzahlbestimmungen für Gruppen, Graphen und chemische Verbindungen, Acta math. **68**, 145–254.

— and B. MEYER:
1949 Sur les symétries des fonctions sphériques de Laplace, C. r. Acad. Sci. (Paris) **228**, 28–30.

RÉDEI, L.:
1947 Das „Schiefe Produkt" in der Gruppentheorie ..., Comment. Math. Helvet. **20**, 225–264.

REIDEMEISTER, K.:
1932a Einführung in die kombinatorische Topologie, Braunschweig.
1932b Knotentheorie, Erg. d. Math. **1**, 1.

ROBB, A. A.:
1936 Geometry of Time and Space, Cambridge.

ROBINSON, G. DE B.:
1931 On the fundamental region of a group, and the family of configurations which arise therefrom, J. London Math. Soc. **6**, 70–75.

SANOV, I. N.:
1940 Lösung des Burnsideschen Problems für den Exponenten 4, Učenie Zapiski Leningrad Univ. **55**, 166–170.
1951 On a certain system of relations in periodic groups with period a power of a prime number, Izvestija Akad. Nauk SSSR, Ser. Mat. **15**, 477–502.

SCHENKMAN, E.:
1954 Two theorems on finitely generated groups, Proc. Amer. Math. Soc. **5**, 497–498.

SCHLÄFLI, L.:
1858 An attempt to determine the twenty-seven lines upon a surface of the third order ..., Quart. J. Math. **2**, 110–120.

SCHLEGEL, V.:
1883 Theorie der homogenen zusammengesetzten Raumgebilde, Verh. K. Leopold.-Carolin. Dtsch. Akad. Naturforsch. **44**, 343–459.

SCHMIDT, O.:
1924 Über Gruppen, deren sämtliche Teiler spezielle Gruppen sind, Recueil Math. Soc. Math. Moscou **3**, 367–372.

SCHOENFLIES, A.:
1891 Krystallsysteme und Krystallstructur, Leipzig.

SCHOUTE, P. H.:
1912 On the characteristic numbers of the polytopes $e_1 e_2 \cdots e_{n-1} S(n+1)$ and $e_1 e_2 \ldots e_{n-1} M_n$, Proc. Fifth Internat. Congress of Mathematicians **2** (Cambridge, 1913, 70–80).

SCHREIER, O.:
1924 Über die Gruppen $A^a B^b = 1$, Abh. Math. Sem. Univ. Hamburg **3**, 167–169.
1927 Die Untergruppen der freien Gruppen, Abh. Math. Sem. Univ. Hamburg **5**, 161–183.

SCORZA, G.:
1942 Gruppi astratti, Rome.

SÉGUIER, J. DE:
1904 Théorie des Groupes Finis, Paris.
SEIFERT, H.:
1932 Lösung der Aufgabe 84, Jber. deutsch. Math.-Verein. **41**, 7—8.
SEIFERT, H., and W. THRELFALL:
1947 Lehrbuch der Topologie, New York.
SENIOR, J. K., and A. C. LUNN:
1934 Determination of the groups of orders 101—161, omitting order 128, Amer. J. Math. **56**, 328—338.
SHEPHARD, G. C.:
1952 Regular complex polytopes, Proc. London Math. Soc. (3), **2**, 82—97.
1953 Unitary groups generated by reflections, Canad. J. Math. **5**, 364—383.
1956 Some problems on finite reflection groups, L'Enseignement Math. (2), **2**, 42—48.
— and J. A. TODD:
1954 Finite unitary reflection groups, Canad. J. Math. **6**, 274—304.
SHERK, F. A.:
1962 A family of regular maps of type {6, 6}, Canad. Math. Bull. **5**, 13—20.
SINKOV, A.:
1935 A set of defining relations for the simple group of order 1092, Bull. Amer. Math. Soc. **41**, 237—240.
1936 The groups determined by the relations $S^l = T^m = (S^{-1}T^{-1}ST)^p = 1$, Duke Math. J. **2**, 74—83.
1937 Necessary and sufficient conditions for generating certain simple groups by two operators of periods two and three, Amer. J. Math. **59**, 67—76.
1938 On generating the simple group $LF(2, 2^N)$ by two operators of periods two and three, Bull. Amer. Math. Soc. **44**, 449—455.
1939 A note on a paper by J. A. Todd, Bull. Amer. Math. Soc. **45**, 762—765.
SOLOMON, L.:
1963 Invariants of finite reflection groups, Nagoya Math. J. **22**, 57—64.
SPEISER, A.:
1924 Theorie der Gruppen von endlicher Ordnung (3. Aufl.), Berlin.
STEINBERG, R.:
1959 Finite reflection groups, Trans. Amer. Math. Soc. **91**, 493—504.
1960 Invariants of finite reflection groups, Canad. J. Math. **12**, 616—618.
STEINHAUS, H.:
1950 Mathematical Snapshots (2nd edition), New York.
STEPHANOS, C.:
1879 Sur les systèmes desmiques de trois tétraèdres, Bull. Sci. Math. (2), **3**, 424—456.
STIEFEL, E.:
1942 Über eine Beziehung zwischen geschlossenen Lieschen Gruppen und diskontinuierlichen Bewegungsgruppen ..., Comment. Math. Helvet. **14**, 350—380.

SUNDAY, J. G.:
1972 Presentations of the groups $SL(2, m)$ and $PSL(2, m)$, Canad. J. Math. **4**, 1129—1131.

THRELFALL, W.:
1932a Gruppenbilder, Abh. sächs. Akad. Wiss. Math.-phys. Kl. **41**, 1—59.
1932b Lösung der Aufgabe 84, Jber. deutsch. Math.-Verein. **41**, 6—7.

THRELFALL, W., and H. SEIFERT:
1931 Topologische Untersuchung der Diskontinuitätsbereiche endlicher Bewegungsgruppen des dreidimensionalen sphärischen Raumes I, Math. Ann. **104**, 1—70.
1933 Topologische Untersuchung der Diskontinuitätsbereiche endlicher Bewegungsgruppen des dreidimensionalen sphärischen Raumes II, Math. Ann. **107**, 543—586.

TIETZE, H :
1910 Einige Bemerkungen über das Problem des Kartenfärbens auf einseitigen Flächen, Jber. deutsch. Math.-Verein. **19**, 155—

TITS, J.:
1962 Groupes simples et geometries associées, Proc. Internat. Conr. Math. 197—221.
1970 Groupes Finis Sporadiques. Seminaire Bourbaki No. 375.

TOBIN, S.:
1960 Simple bounds for Burnside p-groups. Proc. Amer. Math. Soc. **11**, 704—706.

TODD, J. A.:
1931 The groups of symmetries of the regular polytopes, Proc. Cambridge Philos. Soc. **27**, 212—231.
1932a A note on the linear fractional group, J. London Math. Soc. **7**, 195—200.
1932b Polytopes associated with the general cubic surface, J. London Math. Soc. **7**, 200—205.
1936 A second note on the linear fractional group, J. London Math. Soc. **11**, 103—107.
1947 On the simple group of order 25920, Proc. Roy. Soc. London, A **189**, 326—358.
1950 The invariants of a finite collineation group in 5 dimensions, Proc. Cambridge Philos. Soc. **46**, 73—90.
1959 On representations of Mathieu groups as collineation groups, J. London Math. Soc. **34**, 406—416.
1970 Abstract definitions for the Mathieu groups, Quart. J. Math. Oxford (2), **21**, 421—424.

— and H. S. M. COXETER:
1936 A practical method for enumerating cosets of a finite abstract group, Proc. Edinburgh Math. Soc. (2), **5**, 25—34.

TÓTH, *see* FEJES.

TROTT, S.:
1962 A pair of generators for the unimodular group, Canad. Math. Bull. **3**, 245—252.

TROTTER, H.:
1964 A machine program for coset enumeration, Canad. Math. Bull. **7**, 357—368; Program listing in Mathematical Algorithms **1** (1966), 12—18.

VAN DER WAERDEN, *see* WAERDEN.

VEBLEN, O., and J. W. YOUNG:
1918 Projective Geometry **2**, Boston.

VORONOI, G.:
1907 Sur quelques propriétés des formes quadratiques positives parfaits, J. reine angew. Math. **133**, 97–178.
1908 Recherches sur les paralléloèdres primitifs, J. reine angew. Math. **134**, 198–287.

WAERDEN, B. L. VAN DER:
1931 Moderne Algebra **2**, Berlin.
1948 Gruppen von linearen Transformationen, Erg. d. Math. **4**, 2.

WEBER, H.:
1895/1896 Lehrbuch der Algebra, 1 u. 2, Braunschweig.

WELLS, A. F.:
1956 The Third Dimension in Chemistry, Oxford.

WEYL, H.:
1952 Symmetry, Princeton.

WHITNEY, H.:
1933 Planar graphs, Fund. Math. **21**, 73–84.

WITT, E.:
1937 Treue Darstellung Liescher Ringe, J. reine angew. Math. **177**, 152–160.
1938 Die 5-fach transitiven Gruppen von Mathieu, Abh. Math. Sem. Univ. Hamburg **12**, 256–264.
1941 Spiegelungsgruppen und Aufzählung halbeinfacher Liescher Ringe, Abh. Math. Sem. Univ. Hamburg **14**, 289–322.

YOUNG, A.:
1930 On quantitative substitutional analysis, Proc. London Math. Soc. (2), **31**, 273–388.

ZASSENHAUS, H.:
1935 Über transitive Erweiterungen gewisser Gruppen aus Automorphismen endlicher mehrdimensionaler Geometrien, Math. Ann. **111**, 748–759.
1940 Ein Verfahren, jeder endlichen p-Gruppe einen Lie-Ring mit der Charakteristik p zuzuordnen, Abh. Math. Sem. Hamburg **13**, 200–207.
1958 Theory of Groups (2nd ed.). New York.
1969 A presentation of the groups PSL (2, p) with three defining relations, Canad. J. Math. **21**, 310–311.

Index

B. Huppert

Endliche Gruppen I

Nachdruck. 1979. 15 Abbildungen.
XII, 793 Seiten. (Grundlehren der mathematischen Wissenschaften, Band 134)
ISBN 3-540-03825-6

Inhaltsübersicht: Symbolverzeichnis. – Grundlagen. – Permutationsgruppen und lineare Gruppen. – Nilpotente Gruppen und p-Gruppen. – Verlagerung und p-nilpotente Gruppen. – Darstellungstheorie. – Auflösbare Gruppen. – Literaturverzeichnis. – Namenverzeichnis. – Sachverzeichnis. Errata.

Aus den Besprechungen:
"...Keiner, der in die für die heutige Forschung der endlichen Gruppen wichtigen Gebiete eindringen will, wird an diesem ausgezeichneten Buch vorübergehen können. Für die Kenner der Materie wird es zu dem Handbuch werden... Der Verfasser versteht es ausgezeichnet, komplizierteste Vorgänge klar darzustellen. Er benutzt dabei auffallend kurze Sätze (wenig Nebensätze), die den Formulierungen eine ausgesprochene Brillianz geben. Besonders lobenswert sind die jedem Kapitel (mit Ausnahme des ersten) vorangehenden Motivationen, die den Lernenden die Bedeutung des Folgenden ahnen lassen. Dabei wird gleichzeitig ein Ausblick sowie eine Abgrenzung gegen andere Darstellungsmöglichkeiten gegeben. Neben zahlreichen interessanten, keineswegs trivialen Aufgaben erfreuen viele Hinweise auf ungelöste Fragen und auf Ergebnisse, deren Darstellung den Rahmen des Buches sprengen würden. Ausführliche Literaturbemerkungen zu jedem Kapitel und ein umfangreiches Literaturverzeichnis (Stant 1967) ergänzen das Werk..."

Mathematisch-Physikalische Semesterberichte

Springer-Verlag
Berlin
Heidelberg
New York

QA171 .C7 1980
010101 000
Coxeter, H. S. M. (Harold
Generators and relations for d
0 2002 0024402 4
YORK COLLEGE OF PENNSYLVANIA 17403